BUYER'S GUIDE
1965–1998

T0328071

CLASSIC
PORSCHE
911

RANDY LEFFINGWELL

UPDATED BY MATT STONE

motorbooks

**To the members and hard-working officers of the Porsche Club of America,
for their enthusiasm and their efforts at preserving all of Porsche's great road cars.**

Brimming with creative inspiration, how-to
projects, and useful information to enrich your
everyday life, quarto.com is a favorite destination
for those pursuing their interests and passions.

© 2022 Quarto Publishing Group USA Inc.
Text © 2002, 2009, and 2022 Randy Leffingwell

This edition published in 2022
Second edition published in 2009
First Published in 2002 by Motorbooks, an imprint of The Quarto Group,
100 Cummings Center, Suite 265-D, Beverly, MA 01915, USA.
T (978) 282-9590 F (978) 283-2742 Quarto.com

Motorbooks titles are also available at discount for retail, wholesale, promotional, and bulk purchase.
For details, contact the Special Sales Manager by email at specialsales@quarto.com or by mail at The
Quarto Group, Attn: Special Sales Manager, 100 Cummings Center, Suite 265-D, Beverly, MA 01915, USA.

26 25 24 23 22   1 2 3 4 5

ISBN: 978-0-7603-7719-2

Digital edition published in 2022
eISBN: 978-0-7603-7720-8

Originally found under the following Library of Congress Cataloging-in-Publication Data

Leffingwell, Randy, 1948-
Porsche 911 buyer's guide / Randy Leffingwell. – 2nd ed.
p. cm.
ISBN 978-0-7603-3484-3 (sb : alk. paper)
1. Porsche 911 automobile—History.
2. Porsche 911 automobile—Purchasing. I. Title.
TL215.P75L435 2009
629.222'2—dc22
2009004353

Cover Design: Cindy Samargia Laun
Cover Images: Main, Dieter Landenberger; Details, Michael Alan Ross
Page Layout: Tom Heffron and Katie Sonmor

Printed in China

Contents

Preface

It's very difficult to write a useful buyer's guide without pointing out problems and risks. Overall, such a book could sound negative about the subject at hand. But that is not my intent.

My philosophy is "forewarned is forearmed." Not every owner may be as conscientious as you will be when you buy your 1st or 14th Porsche.

Make no mistake: There is nothing like a Porsche. They are long-lived sports cars, not disposable appliances. The two oldest cars in this book are still in the hands of their original buyers. Four cars in this book each have more than 275,000 miles on them, with more than 200,000 of those driven by current owners. They are engineered to be driven, not just aimed or "operated." For nearly 75 years years Porsche owners and drivers have known a truth: Germany's other auto company using the phrase "The ultimate driving machine" has only appropriated that expression from the world's true ultimate driving machine.

One of these owners has a sticker on the rear window of his daily driver 911 Carrera. Written in German, it parodies another carmaker's claim. It translates "Porsche: The Heartbeat of Germany."

The noise, the sensations of accelerating a Porsche to redline, of trail-braking at the edge of adhesion into a turn, of riding that squat, that lateral sway, is unmatched. No other car drives or sounds like this.

Porsches are not for everyone. Consider the following quotes from resellers interviewed for this book: "A 911 Porsche is something YOU have to be in tune with." "You have to check the oil and understand that running the engine harder is better." "Run it to the redline at least once a day."

Porsches have always been expensive. Factoring in inflation, the Gmünd coupe sold for the same price as a new Boxster S today. The designing, engineering, and testing needed to build cars capable of 300,000 miles of high-performance driving requires time and strong parts, and these cost money. Not all Porsches have been perfect from the start. But with each passing year, 911s have got better and stronger.

Go find a good 911. Drive hard. Smile a lot.

—*Randy Leffingwell, Santa Barbara, California*

ACKNOWLEDGMENTS

The minutiae involved in producing a book such as this requires so much help. As far as "technical advisers" to this second edition, I am very grateful to the late Bruce Anderson, Drytown, California; Walt Branscome, Jason Horneman, and John Ramsey, Santa Barbara Auto Group, Santa Barbara, California; Tony Callas and Tom Prine, Callas Rennsport, Torrance, California; Scott Hendry and Dan Reese, Scott's Independent, Inc., Anaheim, California; Henry Hinck and Joe Schneider, Schneider Autohaus, Santa Barbara, California; Bill Raifsnyder, Houston, Texas; and to Pete Stout, *000* magazine. Their insight, information, experience, and time that they shared with me are beyond measure.

Doing any book about Porsche is so much easier with the generous assistance from Porsche Cars North America. In particular, I'm grateful to the late Bob Carlson, manager, motorsport and brand heritage; Bernd Harling, general manager, public relations; Robin Baker, administrative assistant; Gary Fong, press fleet manager; and Dave Engelman, product communications.

I further wish to acknowledge the late Patrick Paternie and Peter Bodensteiner for their detailed and important *Porsche 911 Red Book*. Any holes discovered in this buyer's guide can be filled very capably by their *Red Book*.

The logistics of locating and photographing these automobiles required extensive help and cooperation. My sincere thanks go to: Joel Adler, Camarillo, California; James Alton III, San Dimas, California; Howard Barker, Pacifica, California; Oscar Briones, Santa Barbara, California; Dennis Carpenter, Camarillo, California; Jack Croul, Corona del Mar, California; Douglas Dodge, Aliso Viejo, California; Frank Enea, Monterey, California; Kenn Funk, Monrovia, California; Sam Gallucci, Westlake Village, California; Gary Griffiths, Santa Cruz, California; Marty Harris, Simi Valley, California; Scott Hendy, Anaheim, California; Rob Heyne, Arroyo Grande, California; Hal Holleman, Newport Beach, California; David Jacobs, Ventura, California; Ray Jordan, Santa Barbara, California; Ed Justice Jr., Justice Brothers, Inc., Duarte, California; Leon Kreger, Pebble Beach, California; Pete Lech, Fullerton, California; Jesse Lieber, Santa Barbara, California; Frank Luz, Long Beach, California; Jim Middlebrook, Santa Rosa Valley, California; Bill Nakasone, Newport Beach, California; Charles Noble, Santa Ana, California; Loren Peters, Palos Verdes Estates, California; Les Quam, Las Vegas, Nevada; Dan Reese, Anaheim, California; Ben Rodriguez, Ventura, California; Fred Rosenbloom, Ontario, California; Chris Roman, San Francisco, California; Ed Scheid, Corona del Mar, California; Jim Schrager, Mishawaka, Indiana; Fred Stewart, Burbank, California; Kathy Tasaka, Pasadena, California; Stewart Thomas, Tustin, California; George Vorgitch, Oxnard, California; Chester Yabitsu, Agoura Hills, California; and Jay Yard, Brentwood, California.

To you all I offer my deepest gratitude.

Introduction

For years now, after taking over production of the *Illustrated Porsche Buyer's Guide* from my friend the late Dean Batchelor, numerous friends have called me seeking advice on which Porsche to buy. That led to the *Porsche 911 Buyer's Guide.* Since publication of the first edition in 2002, times and Porsche as a marque have evolved. There are many flavors of 911ophile, yet they most commonly break into the largest groups defined by "early" or "classic" air/oil cooled models, and the later "modern" water-cooled generations. So we've redeveloped this book to concentrate on the former from the 911's birth in the mid-1960s to the final air-cooled 993 models of 1998.

The opening chapter still provides you with a detailed procedure to follow when you go shopping for a car. It is oriented toward Porsche 911 models, but you can use the general recommendations and procedures to shop for a Pontiac or a Packard as well as your dream 911 Turbo. The second chapter addresses the open cars, the Targa and Cabrio models, for those buyers interested in these cars. Chapter 3 introduces you to the murky world of gray-market cars.

You will likely notice that certain commentary and information appears to be repeated – this is a natural byproduct of the fact that certain comments and caveats carry over from model to model, or year to year. Other notes may disappear from later entries as those issues were addressed or resolved as the 911 evolved.

Each chapter contains a box that reports prices for an identical list of parts and service from one model year to the next. Labor prices for minor and major service at dealerships and independent shops vary across the country, but generally range from $110 to $200 per hour at this writing. I've figured labor at an average of $150 per hour.

Parts prices quoted for new factory parts are list prices, in US dollars, unless otherwise noted. For those interested in a car whose parts originate in Euro prices; the exchange rate is a difficult issue to consistently gauge. The value of the Euro against the dollar is a constantly varying barometer of the world economy. Consider these figures as a rough estimate accurate at the time of publication. So I beg your indulgence: Prices, as listed here, are "accurate" as this book goes to the printer. I recommend you ask your local shop or dealer what to expect for their service charges. Prices quoted here are for parts only and do not include the labor to install them. I have chosen parts for normally aspirated, two-wheel-drive 911 Coupes, unless otherwise described. Generally prices for the Targa, Cabriolet, Turbo, or Sport models will be higher. I've listed "used" and "restoration" prices for items no longer available (NLA) from the factory. Nonfactory (aftermarket) replacements are available for nearly all the items listed, often at reduced cost. However, the quality can vary greatly.

This book does not examine racing models. Among collectors, these cars change hands quickly and for great sums. With this book, you can read about and make your own judgments on regular production street models, such as you can find on websites, bulletin boards, by word of mouth, in print media, and at auto sales lots and auctions. Some years and some models have earned decidedly bad reputations. Most definitely there are even more cars to desire with every molecule of your enthusiast being.

The goal of this book is still to make your shopping, examining, and purchasing stresses manageably low, so your driving and pleasure quotients remain very high. Look carefully, do your homework, and be patient while you shop. Be ready to walk away from the wrong car, and be ready—with your finances in order—when you find the right one. Be patient. The car you desire is out there.

A NOTE ON RATING

I considered these ratings carefully. They reflect input from and the consensus of many sources. If you disagree with any of these evaluations, please look to the text for explanations.

Acceleration, comfort, and handling are the areas in which the car is compared to its competition when it was in production. "Parts availability" and "reliability" refer to the present day, in practical terms of the car being a daily driver. It's important to remember that long-term reliability is only as good as the quality of the mechanic who performed the repair.

The goal of these ratings is to get you only into the ballpark of your car selection. To get to home plate, you must read the text and the Garage Watch entries.

What the ratings mean:

5 – Equal to or better than the rest of its competition.
4 – Almost equal to the best, better than most others.
3 – Comparable to its competitors. Not bad, just average.
2 – Below average or expectations. See text.
1 – Don't leave home without a cellular phone.

Chapter 1
How to Buy
a Used Porsche

How do you find a good 911?

The only way to be certain you're buying a solid car is to take it to an authorized dealer or qualified independent service facility for a prepurchase inspection. Make an appointment and expect to pay between $300 and $750 for this examination. A Porsche dealer service technician will provide you a printed list of problems and estimated repair charges for anything they find. You can take this list back to the seller and use it to adjust the selling price accordingly.

But what if you can't afford a dealer inspection for every car you like? And what if there's a car on a lot whose owners won't let you take it for an inspection, yet the car seems solid and strong? Without an inspection, you do take a risk.

You can check some things yourself to eliminate flawed candidates. Throughout this chapter I'll advise you to thank the seller for his or her time and walk away. These problems are serious enough to cost you real money. Unless you are an experienced Porsche mechanic able to do the work yourself, just thank the seller and walk away.

For this screening process, you'll need no tools other than your eyes, ears, and nose. You will get your hands dirty once; a Wet & Dry paper towelette will clean you up. You may want a thermometer to stick into the air vent to check air conditioning and heating temperature. A tire gauge will be beneficial, too. You might bring a flashlight to shine into the engine compartment and a small mirror to hold in places where you cannot get your eyes. An observant friend accompanying you can serve as a second pair of eyes and, equally important, can occupy the seller while you examine the car.

Don't be open with the seller. Your color preference and your budget should remain your secret. Don't buy a car at night; what you can't see, you'll pay for later.

These tests will eliminate some dangerous and costly problems. As the car passes each stage, you'll eliminate risks of what may be wrong, and you'll understand what is right with the car you're looking at. You can be more certain that you're getting a better car.

A bit of advice here: Don't drive only one of the models you want. Drive at least three, especially if one is beyond your budget. You must build a log of experiences that you have to draw on for that moment when you find the right one at your price.

SECTION 1. BEGIN WITH THE CAR BODY

1A. Run your fingers along body seams, door panels, and doorjambs and around the roof. There should be no uneven surfaces, gaps, or seams. Where fixed windows (and windshield) seals attach to the body, check for gaps and undulations. This might mean the car has had a big accident. Look for paint that appears bubbled below the surface. This indicates rust, something that plagued early 911s until the factory began galvanizing the steel to protect it. But some later models rust as well.

1B. View the entire car from all angles. Use open shade to do this. Direct sun conceals more than it reveals.

1C. Look at repair records for the life of the car. Be wary if a private owner has no receipts. (Some owners drive Porsches for partial business use and need to keep their receipts; you don't need to have them, just look them over.) Used car lots seldom have the records. Authorized dealers can check their computers for service histories but these will not always show body damage.

1D. Everyone consulted for this book agreed you should not buy a car that has crashed. Porsche 911s are unitized construction, meaning that it has no frame as such. The floor pan, other structural members, and many body panels constitute the "frame." The worst case is a car the insurance company declared a total loss. Then a body shop acquired the wreck as "salvage" and rebuilt it. Yet even this is not a black-and-white rule. One car in this book was a "salvage," and it runs straight. Its title gives it away, and its resale value always will reflect that as a result.

Theft recovery is another kind of "salvage." Typically enough of the car has been removed that the insurance company determines it is cost-effective to replace the car. These cars generally have not been damaged, just stripped of resalable parts. It will cost you the price of the car plus the parts. If you are a competent mechanic, this may be a good bet; otherwise you'll pay someone else a great deal to reinstall what was stolen.

1E. Generally leased vehicles are adequately maintained. But lessees, knowing the car isn't their own, often drive hard and arrive late for required service intervals. Service records for cars done at U.S. authorized dealers are computerized and retrievable.

1F. Inspect the car's service manual to see that all warranty period service intervals were met and performed by authorized Porsche dealers. If there are no entries at all in the factory-provided service book, thank the seller and leave now.

SECTION 2. LOOK AT THE ENGINE

2A. Shine your flashlight onto every engine surface you can see from above and below. The small mirror will show you the back and undersides. Look for gross evidence of leaks, big smears of oil, or obvious trails of gasoline. Slight seepage is acceptable. If the rest of the car checks out through this screening process, the mechanic who does the prepurchase inspection will notice the seeps and tell you what they signify.

2B. With the engine compartment open, watch the engine and exhaust pipes.

Have the seller start the engine. Does it rock back and forth as the starter engages, suggesting the motor mounts have failed? (That's a $1.000 repair, including parts and installation.) Does it take long to start? Is what comes out of the exhaust pipe white, gray, black, or invisible?

On cool or cold mornings, some pale gray steam is normal. As the engine warms, this should stop. A slight puff of white smoke, especially on earlier models, is entirely acceptable. A steady cloud suggests the engine burns oil. Smell the exhaust. If it smells like oil, thank the owner and leave.

Black smoke may come from a vacuum leak, emission system failures, or fuel injection system malfunction. Each is costly to track down and repair. Black smoke also may indicate an engine loaded up with carbon because its driving use is short, in-town jaunts. The engine never gets truly warm, so fuel and oil deposits accumulate in the engine.

2C. Do not touch the gas pedal when starting a fuel-injected car. If the seller must pump the gas pedal on a fuel-injected engine, this indicates a massive vacuum/air leak that prevents the engine from getting enough fuel to start on its own. This is an expensive problem. In addition, the engine will not pass smog tests in most states. It may burn exhaust valves and perhaps pistons. Shut off the engine (if it even starts), thank the seller, and leave.

NOTE: FOR YOUR SAFETY, make sure the engine hood stays up. Struts are $90 each, installed, and the seller might not have replaced those. On the Turbo models, these hold up an oil cooler, rear window wiper and motor, and the heavy rear wing as well.

2D. Listen for any noise besides a nice exhaust sound, such as tapping, thumping, whirring, or hissing. Let the engine run for

These 1965 models represent the 911 in its first production year and the 356 C in its final year. For 1948, Erwin Komenda's organic-shaped 356 was as startling as Butzi Porsche's angular 911 25 years later.

10 minutes and check temperatures. The needle should remain below the halfway point on the oil temperature gauge and (on 996 and 997 water-cooled models) the water temperature gauge.

2E. Smell for oil, coolant, or gas. If you find any of these, shut the engine off immediately. Some models develop engine compartment gas line leaks that can start a fire.

2F. If you didn't smell these critical fluids while the engine was running, turn it off, sniff again, and listen carefully. Crackling sounds coming from the engine of the water-cooled cars suggest the engine is overheating in the cylinders. Water jackets may be clogged, oil may not be circulating properly, the oil cooler may be clogged, and/or the pump may be failing.

Gurgling sounds suggest problems in the cooling system, such as insufficient flow through the radiator or a failing water pump on 996 and 997 models. The oil cooler or radiator may need cleaning or replacement. The oil or water pump may need replacement.

2G. With the engine off and warm, check the oil hoses, not before it's started. Don't squeeze with your fingers; they're hot. Use the eraser-end of a pencil or the flat end of a ballpoint pen to press the hoses. If the hoses are firm when they're hot, they are fine. If the hose is too soft, the inner lining has broken down and the hose will fail at some time soon. If the hose has swollen around the clamps or fittings, this also indicates hose failure.

SECTION 3. TIRES, WHEELS, BRAKE ROTORS, AND SUSPENSION

Before you move the car, check the air pressure in all four tires. This will give you the "cold" pressure, and it should be within the ranges you'll find on the chart below. This also will give you an indication of the seller's attention to routine maintenance; a slight

Besides the more specialty limited editions such as the Carrera RS and the 911R, the early gen (1967–1973) S model has become the darling among "chrome bumpered" air-cooled 911s; it's the raciest among them, offering the most power, and the mechanical fuel injection that so many ardent enthusiasts want. Good ones cost big money these days.

Few buyers will remove a headlight to check for rust, but be aware that many cosmetic repaints camouflage corrosion behind good looks. If you are examining a car that has lived all its life near an ocean coast or in northern states that use road salt in winter, this inspection may be worthwhile.

difference is reasonable, but differences of more than, say, 5 or 8 psi from one tire to the next hints at casual care, not something desirable in a preowned Porsche. (If tire pressures are off or low, make a note of cold pressures, and when you stop at a pump before going on the driving test, adjust pressures accordingly. More on this in the driving section.) This is necessary to determine if handling ills are tire pressure related or represent much more serious problems.

You'll get your hands dirty now.

3A. Rub your hand along the tread on all four tires, feeling for ridges or undulations. Generally someone selling their car won't spend $1,000 or more to replace tires, so these will tell you a great deal. Be sure they have equal tread depth, are the same brand, and are the appropriate size for the car. Mismatched tires suggest the seller dealt cheaply with other service needs as well.

Ridges on treads indicate poor shock absorber condition. Undulations suggest the alignment is wrong or suspension is damaged, from hitting a curb, for example.

Examine all four disc brake rotors. Run your finger along the surface of the rotor. You should feel no ridges, just smooth, even surfaces. Ridges indicate the owner has run the pads past tolerance. Grooves (which you can feel with your finger tips) mean the owner used aftermarket brake pads that are riveted together. These wore beyond tolerance and scored the rotor. Factory pads are bonded and use no rivets for just this reason.

Feel the outside edge of the rotor. If there is a lip, it suggests the owner has ignored all service warnings that the pads have gone too far and have begun to score the rotors. Depending on the depth of the lip, the rotor may be beyond tolerance. Some people say you can turn a Porsche rotor once. Others say no. Thank the owner. Leave now.

This is the ideal way to encounter Porsche's Typ 905 Sportomatic transmission: after a complete rebuild, sitting on a work bench, on display, not intended for use. It was not a bad transmission, but now parts and, especially, seals are spectacularly expensive. Owners often replace ailing Sportomatics with manual transmissions.

The 2.2-liter engine decals are no longer available from Porsche, but they are available from several aftermarket sources. They are desirable for restorations.

If there was no lip, reach inside the wheel wells to feel the inside edge of the wheels. Porsche wheels, forged from a solid billet, are the strongest available for the car and among the strongest anywhere. Nearly all aftermarket wheels are cast, poured into molds. Some aftermarket wheels shatter upon impact with a curb. Others distort and can never be trued. If you feel scuffs, bumps, ripples or cuts on the inside, you may have a bad wheel. Later on, you will be able to determine better what this problem may be during an early portion of the road test.

You can clean your hands now.

Now go to the front of the car.

3B. Give a quick, sharp push down on the front bumper. (Not the fender; you can dent it.) Did you feel any binding or hear any noise except a gentle hiss?

3C. As the suspension brings the car back up, does the car clunk or gurgle, or bounce more than once? Does the car rock? It should do none of these. Repeat this on all four corners of the car, making sure that each rebounds (comes back up) at the same rate.

If the suspension does not perform smoothly and quietly, thank the seller and leave. If however, the car has passed these tests, you can go on and look inside the passenger compartment.

SECTION 4. IN THE INTERIOR, EVIDENCE OF CARE OR ABUSE WILL BE MORE SUBTLE

4A. As you swing open the driver's door, feel the door in your hand. Does it bind, squeak, or fall as you move it? It should be even, level, and fluid. Look at the driver's seat around the side edge of the seat cushion or side bolster. If it's cracked or worn, the stitching broken, the inner padding misshapen or lumpy, the car saw mostly short trips around town with frequent driver ingress-egress. Because of this kind of use, the engine rarely reached operating temperature or remained there long enough for the oil

to do its true cleaning functions of carrying deposits and debris to the filter. Short hops actually force these deposits onto engine surfaces and bond them into the varnishlike sludge.

4B. Are the door seals cracked, chipped, or split? When rain drips through these onto the carpet, the carpet soaks it up. The top layer of the carpet dries; the underlayer holds moisture and rusts out the rocker beam, the main structural element whose integrity makes you feel secure pitching the car into a 50-mile-per-hour turn at 80. Lift carpets and mats and look for rust. Crawl underneath and look again. Use your mirror to see where you cannot fit.

Now climb into the car. Just sit. Do not touch anything.

4C. Close your eyes. Move your body around in the seat and feel if the seat moves, wiggles, or makes noise. If so, the floor pan may have been weakened due to a crash or because of rust, which allows play in the seat bolts that should never exist. (This is the reason to check the integrity of the seat belts as well. They are an element of the handling and cornering security you feel.) Another possibility is that the previous owner was extremely heavy individual who enjoyed vigorous driving that has stressed the seat frame and mounting bolts. If the seat does wiggle, thank the owner and leave now. Replacing a floor pan is an expense you undertake only if you're completely restoring a car worth the expense. Given today's demand and increased 911 values, this expensive repair is likely worth doing on nearly any example, assuming the damage isn't too extensive and that the repair bid is fair.

Now look at the general condition of the interior. Is the leather or vinyl cracked or dried out? Are knobs cracked or missing?

NOTE: Leather. Porsche's largest market in the world is Southern California with about 25 percent of its entire production ending up in a hot, sunny seven-county area. Porsche admits it has difficulty with leather in the United States, especially on dash panels. From the mid-1980s through the mid-1990s, Porsche did not offer leather dashboards because their warranty failure rate was so high.

4D. Hold the steering wheel. Try to pull it from side to side and up and down. There should be no play, no movement. Many drivers use steering wheels as handles as they get in or out. Do this enough, you loosen up steering column mounts.

Now adjust the seat to fit your comfort and safety level, making certain you can push the clutch pedal to the floor. (You're not driving yet, but you need the leverage.) Put on the belt. Check it for wear, tear, grease, or any slack and play when it is supposed to be holding tight. Fasten and release it a half-dozen times to make sure the clasps will not hang up when you need to get out of it. Then restart the car yourself without touching the steering wheel.

SECTION 5. TESTING BRAKE PEDAL CHARACTERISTICS AND FIREWALL INTEGRITY

Do this next section with the engine running. You need engine vacuum boost for the power brake system if it's a model year 1977 or later U.S. car.

5A. With your hands in your lap, tap on the brake pedal a couple of times to bring the fluid up to pressure. Then press as hard and as far down on the pedal as you can 10 times. Use both legs.

Pedal pressure must remain constant. If the pedal pumps up and gets harder as you pump it, you can shut off the engine, thank the seller, and go home. A rising brake pedal indicates system problems that may include vacuum leaks, water in the brake fluid, or a failed brake booster.

5B. If the brakes pass this test, keep pumping as hard as possible. What you are doing now is looking for any movement on the dashboard, and listening for creaks and moans.

If there is any noise or movement around the floorboards or in the dash, realize that you may have a serious problem. Chassis integrity may have been compromised, possibly through an earlier bad crash. Equally likely in Porsches, a previous driver was very hard on brakes. They can actually crack the firewall pretending they're Ricky Racer, and every corner is turn one. Or the car saw a lot of time on racetracks, and the owner really was Ricky Racer.

The 1970 and 1971 S models, especially, are rare, making them expensive. But they're an acquired taste: Their 200-horsepower engines developed power high up in the engine speed range, making them difficult for routine city traffic. This example boasted both a factory sunroof and factory air conditioning.

One of Porsche's most desirable models is this lightweight version of the 1973 RS Carrera. It is so popular that many Porsche owners have created replicas that should sell for about one-third of the price of a real example. Check serial numbers on any RS Carrera before you buy. While this model is real, many owners enjoy the replicas because they can drive them anywhere without risking their large investment.

5C. For Sportomatic models from 1968 through 1979 and cars with Tiptronic transmissions, starting in model year 1990, once the brakes and the firewall/dashboard pass this test, keep your feet on the brake and slowly run the automatic gearshift through the gear range. Wait at each selection for a full minute. You should see a drop of 100 to 200 rpm on the tach and feel a very slight nudge as gears engage.

If it takes 30 seconds, for example, for the transmission to catch up to your selection, the warm-up regulator, which controls the vacuum between the engine and transmission, is beginning to fail. Gear engagement will "clunk" in, hard, shaking the car. If you do experience any delay, thank the seller and go home.

NOTE: I strongly advise anyone considering acquiring a car with the Sportomatic transmission to reconsider. Spare parts are difficult to find and costly when you can. Seals can dry out long before they are ever installed and so the cost of the rebuild is wasted when the freshly redone transmission leaks within days. Many owners give up in despair and install manual gearboxes. This conversion can be less costly than a Sporto rebuild.

5D. If the Sportomatic or Tiptronic passes the transmission test, hold your feet on the brake, place the gear selector in second, with the engine at idle. Pull up on the hand brake/parking brake/emergency brake. It should be no more than six to eight clicks with increasing pressure. Slowly release pressure on the

foot brake pedal and be certain the car does not move. If the car moves or if it requires any more than six-to-eight clicks, thank the seller and go home.

(You can do this same test for a manual transmission model by easing in the clutch in first or reverse gears. Look around very carefully before you try this test to make sure you have plenty of open space if the brake fails.)

If you get 15 clicks, for example, or if the car moves, the rear brake system is worn out, and you may need to replace both rear rotors, calipers, and the emergency brake cable. (This can be an expensive job; in some models the cable runs through a tunnel and technicians justifiably charge for their time.)

5E. If everything passes to this point, then, with the engine still at idle, put the car in neutral, parking brake on. When the temperature gauge moves off the bottom stop, and only then, give the gas pedal a couple of good raps.

Testing this before the engine has reached temperature can mask a sticking throttle cable, which you can tell only once engine temperature has begun to rise. The cables frequently stick if the engine has been steam cleaned. The steam cleaner dissolves all the grease and oil on the engine, good and bad. It usually, however, leaves a residue in the ball-and-cap at the end of the throttle cable at the carburetor linkage or fuel injection throttle body. This residue can make the throttle cable stick.

11

ABOVE
This is one of several places, just ahead of the rear tires, to check for rust on early 911 models. If the car has been hit hard, surface rust may be visible around the edges of this trailing arm mount.

Watch for high-pressure fuel leaks out of aging flexible fuel lines. Watch also for failing deck lid struts, the shock absorbers that hold open front trunk, and engine compartment lids. Make sure the lids stay up before you lean in to look.

affect how much air you add. For example, if the low tire reads 20 psi while cold but has increased to 22 psi during the drive, that 2 psi difference will affect how much air you add in this way: If the tire specifications call for 32 psi cold, you should inflate this tire to 34 psi for the remainder of your driving test. Do not, however, overinflate the tire beyond what the manufacturer recommends. That data, maximum psi, is engraved into the tire and can be read along the wheel rim.

(While that makes common sense, it merits a reminder. In very hot climates, such as Arizona or Southern California during the summer, a very low-pressure tire may become very hot if the drive to a service station is some distance at medium or high speed. For example, the 20 psi cold pressure may heat up and rise to 27 or 28 by the time you reach the air pump. Adding those 7 or 8 psi to the 32 psi recommendation may give you 40 psi, a figure that may exceed the manufacturer's recommendation. In that case, go no higher than the maximum and adjust all other tires to match this pressure.) Now you are ready for the remainder of the road test.

Never road test a car in heavy traffic. You cannot possibly get an accurate impression of the vehicle if you are paying close attention to many other drivers.

6A. During the drive, ask the seller to remain quiet. You need to hear the car.

6B. First drive to a smooth stretch of road with a lot of room. Weekends in industrial complexes are best for this. You are testing brakes here. These are not panic stops, just routine braking. At 30 miles per hour, brake with a steady even pressure. You are looking for any undulation in the steering wheel, shaking, pulling from side to side, pulsing or vibrating through the pedal, the noise of metal-on-metal from failed pads, or a chirping sound that indicates cracked brake rotors (around $300 to $400 each installed, not including calipers and pads).

6C. Now find a pot-holed road or broken road surface. You are listening with your ears and feeling with your fingers on the steering wheel as well as the seat of your pants for clunks, rattles, shakes, shimmies, or any unsettling response to potholes or bumps. If doors, side windows, or the large back window rattles, it may be only tired weather seals, or it could be a badly repaired body after a crash. If the car sounds like it's falling apart, thank the seller, etc.

6D. In a large empty parking lot, drive in large circles. With your hands at 10 and 2, turn the wheel so one hand or the other (you'll do both directions) is at 12. Drive safely but fast enough to listen for slight tire squeal on all four wheels. Specifically, on Typ 964 models beginning with the 1989 1/2 Carrera 4 and 1990 Carrera 2 series, the rear wheels are adjustable for camber and toe-in. Any squeal on any model from front or rear may mean tire pressure is low. If the rear tire pressure is right, a squeal, especially on 1989 1/2 and later models indicates alignment maladjustment or very costly suspension problems.

5F. Once you are off the peg, the engine rpm response should be instantaneous with a nice growl from the exhaust. Now press the pedal and raise the rpm to 2,500 and release it while watching the tachometer needle. The tach, beginning in 1976, is electronically actuated and spring loaded. The rpm needle should fall back in sync with the engine noise. There should be no needle lag. A lag signifies moisture in the electronics. If there is lag, thank the seller and go home.

SECTION 6. NOW IT'S TIME TO GO ON THE ROAD TEST

NOTE: Wearing seat belts is the law. Insist that anyone who may ride with you on any portion of this test use both seat and shoulder belts.

If the air pressure was low on any or all of the tires, your first stop on the road test should be a service station. During the drive to the station, the tires will have heated up slightly. Measure the pressure again in the low tire, noting the difference between cold pressure and current pressure. That heated up difference will

Here is an uncommon model: the 1976 930 Turbo Carrera. While modern Turbos provide more than double the horsepower, they are actually easier to drive than the early cars because Porsche has so greatly improved handling and roadholding. A number of the 1976 Turbos were spun when the turbo boost hit mid-corner, or the driver came off the gas abruptly in a turn. Early Turbos required a very high level of driver skill.

Most visitors to a 911 engine compartment will recognize the air conditioner compressor at right of the cooling fan, but few will identify the air pump on the left. The owner of this 1977 911S, in order to meet California smog requirements, had to invest more than $3,000 on a completely equipped 1978 3-liter engine in order to license the car in California. The previous owner had removed the emissions equipment from the original engine.

ABOVE

Color is a matter of taste and choice. While this Copper Brown Metallic is the original, at least four cars in this book wear nonoriginal colors to satisfy current owners. Other than on historically significant examples such as the first 911 produced, or the 500,000th built, and provided the body is well prepared, there is no reason a car you find cannot be redone if everything else is good.

This faded paint was originally Guard's Red, Porsche's most popular choice in the United States. This is a good indication of a car that lived outside of a garage. The clear coat is "lifting," that is oxidizing and pulling away from the color below it. In the past, this meant a lower price but that is no longer the case today.

Steering should feel smooth with no chatter in the steering wheel. If it tends to move in the direction of your turn, this is called "falling" and can indicate bent struts.

6E. If you have a bent front wheel, you'll find this now. Turning in one direction, if the car shakes or shimmies, the wheel that is bent will be the one outside of the turn. Confirm this by turning in the opposite direction. This unloads the possibly bent wheel and the car should track and ride smoothly. Reconfirm this by resuming the original turn direction. If the car again begins shaking, you have a bent wheel on the outside. Bent wheels most often are on the curb side, inside the wheel where you cannot see it. Porsche wheels are expensive. Negotiate this in your purchase price.

6F. Your dealer or independent shop will check on-board computer systems and determine if the anti-lock braking system (ABS, introduced in model year 1989 1/2 with the Typ 964 Carrera 4) is functioning properly. If you've never stopped with ABS before, the slowing sensation will feel like you've driven into glue. You will feel the brake pedal push back and vibrate against your foot as you hear a rapid, noisy clicking under the dash. The pedal pressure feedback can be unsettling. The clicking is the unit cycling the brakes on, off, and on again faster than you can. You may

Scott Hendry, an independent shop owner in Anaheim, California, says, "Every 911 is at least a $100,000 car, no matter what you pay for it. The least expensive car you can get is the one you'll pay the highest price for. Any S will be $150,000 by the time you get it sorted out. Any G-series will be $100,000. An engine rebuild is $50,000." Don't run from a car like this, however, until you know if its engine is tight, its handling crisp. Then you'll figure out if you want to paint it.

hear rapid tire chirps as the tires near lockup and then release. Anti-lock braking system failure is nearly nonexistent in Porsches.

During your braking test, if the car has done anything other than stop in a straight line in a very quick manner, thank the seller, etc. If the car does stop well with no muss or fuss, then continue on to the next phase.

6G. Now, find a quiet alley with walls on both sides to give you the auditory feedback you want here. Drive only at idle speed with both windows open listening for any noise other than exhaust and tires. If you hear clunking, whirring, tinking, knocking, squeaking, or any other noise besides exhaust purr and tires rolling, thank the . . . . You know the drill by now.

SECTION 7. IF THE CAR PASSES ALL THIS, GO TO A FREEWAY

This is NOT a high-speed test, just a freeway on-ramp merge test.

7A. You want a long freeway entrance ramp where you have great distant vision. You are not seeking record 0–60 times but only impressions. You do not want a speeding ticket.

7B. The acceleration must be impressive and smooth, except for the surge that occurs when the engine comes on the power curve, or "on the cam," or when, at about 3,500 rpm, the turbocharger becomes effective. There should be no bucking or surging.

7C. Porsches since 1989 1/2, whether automatic transmission or stick shift, have the same performance. This is a Porsche selling point. If any 911 model does not seem fast, the engine may need an overhaul. This can easily exceed $10,000 to return an ailing engine to its original performance, a huge concern if you are looking at a car selling for less than that.

If the car you drive is not fast, if you do not get a favorable impression of its performance, thank the seller . . . .

NOTE: Exhaust emissions are bigger problems in some states than in others. Check specific requirements in your state before purchasing any automobile. In California, for example, a successful smog test is a condition of purchase; the completed certificate accompanies transfer of title. The seller pays for this

Between model years 1980 and 1982, Porsche joined the rest of the world in providing energy-conscious Americans with speedometers that read only to 85 miles per hour, since our highways allowed only 55 miles per hour maximum speeds. That the cars still were capable of over 140 miles per hour never seemed ludicrous to anyone in authority.

test in California. This and other requirements and responsibilities differ from state to state. In states such as New York, the vehicle inspection is all-powerful. Whatever your state's standard is, your car must meet it or you cannot get it financed, insured, or licensed.

So, you made it? The car went forward, arrow true, it stopped as if the Hand of God reached down and grabbed you. It made only noises it was meant to do. Well, you've probably found a good one. But you're not done. Return to the seller's driveway, because now is the time for the really important stuff.

SECTION 8. THE BENEFITS OF LIVING IN A CIVILIZED WORLD

Perform this last set of evaluations with the engine idling, the transmission in neutral, and the hand brake on.

8A. Check out the radio. Check all bands and bring a cassette and a favorite CD with you. Make sure it all works and that it will eject your cassette and/or CD without drama and, if it has a power antenna, raise and lower it at least three times checking for jerking or loud whirring noises. Older systems and some well-used new ones fail to eject tapes or CDs; antennas might not reach maximum extension or retraction after a second run.

8B. Turn on the air conditioning to maximum cold and wait. If leaves or debris blow out the vents, the drip tray near the windshield has cracked, allowing rain or car wash water to accumulate in the drip tray, backing up into the vents. This can flow into the cockpit through these vents. Routine factory services calls for cleaning these drip trays but often this gets missed. Wet leaves can mildew inside the car.

Turn the fan back through all speed ranges, waiting long enough to be sure they work. Then back to maximum and insert your turkey thermometer into the air vent. It should read between 32 and 38 degrees (factory specs). If not, the system at least needs an evacuation, oiling, and recharging. When you take it in for this work, insist the shop do both the evacuation and oiling. A shortcut here can cost you thousands later.

This is what you hope to find when you answer an ad. Meticulously maintained, with nearly every piece of paperwork from new, this 1983 also features the optional front and rear spoiler. Still, even with a car like this, a prepurchase inspection is an excellent investment to know what you can expect in the future.

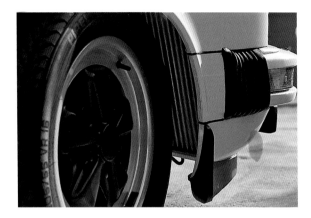

Credit card companies may say "never leave home without them," but Porsche mechanics advise don't leave home without an auxiliary oil cooler such as this one, cleaned of its corrosion proofing and correctly mounted in the front wheelwell. Not only is this more visually pleasing, it allows more airflow to better cool the oil.

The hard fuel lines across the top of the compartment leading down around the fan shroud represented a vast improvement over softer lines on earlier models. Porsche retained flexible lines between the car body and the engine, and these are beginning to wear from age and prolonged exposure to heat and ozone.

Back at maximum cool, watch the oil temperature gauge. It will go up. How fast and how high and what happens next is what you are watching. Air-cooled 911s, beginning with model year 1987 Carrera 3.2 models, came standard with a thermostatically controlled front fan for the oil cooler. Many 911 owners already have installed front oil coolers and additional cooling fans on older models. If the 1973 or newer car you are looking at does not have these, you should plan on doing this. It will save your engine in hot weather and heavy traffic.

If the temperature gauge does not come back down, a sensor, sending unit, thermo-switch, or fan motor may be out.

8C. To reduce engine temperature, and for your next test, switch off the air conditioning and turn the heat and defrost through all speeds to maximum fan speed and maximum heat. Do this even on a 110-degree day or you may find on a 40-degree night that you have a failed heater core, temperature regulator, or heater/defroster fan. These are costly, both for parts and labor.

As you begin to bake, and you will from model year 1977 on when Porsche added a heater fan blower, watch the temperature gauge. It should drop slightly. Be alert for hot oil or radiator coolant or other baking smells inside the car. This could signify a heater core failure or other problems. To make this slightly less challenging for your physical comfort—but even more of a test for the car's electrical system—you can do this portion while checking other important things:

8D. Operate the sunroof. Open and close it at least two times. Do this while the heater/defroster fan is on full. Also switch the rear window defogger on at this point. The sunroof motor develops a high electric current draw, as does the heater fan. This will let in some cool air as you test the electrical system and fuse panel while loading the alternator. It should not blow a fuse. If it does, say thanks and goodbye.

If the sunroof motor whines (sounding like a reeeha-reeeha-reeeha noise), this means the nylon rails or nylon worm drive wheels are failing. Check the weatherstripping around the open roof looking for tears or car wax buildup. Either of these will allow moisture in. If the roof fails to open or close, it may be anything from a fuse, to a motor, to the switch, or wiring, and may be a very reasonable expense or not.

While you're watching the sunroof, look for headliner tears, warped sun visors, stains, or other discoloration on the liner or visors. This indicates moisture already has gotten in. If the sunroof fails to operate or stops while you're testing it, thank the . . . . Oh, you know.

8E. If you are inspecting a model year 1987 or later Cabriolet with a power top mechanism, ask the seller to operate the top through at least two full open-and-close cycles. (If it's a 1983 to 1986 Cabriolet, a 1989 Speedster, or a 1994 Speedster, ask the seller to demonstrate the manual procedure to make sure everything works as it should.) Do this only on a flat, level surface. Make certain every function of the operation is working and that windows raise and lower as they are meant to on later model cars with power roofs. There are specific things to watch for during this test. See the Cabrio sections for important details.

8F. If you are looking at the Targa models, two important facts to know are that they can leak and that Porsche has no more replacement roofs available. They did not produce extras. However some independent shops get them after body shops remove Targa bars to convert the cars to Cabriolets. Targa roofs can be reconditioned. Still, if the panel you are examining is badly damaged or is missing, thank the seller and leave.

Porsche's own air conditioning vents prove inadequate in some regions, such as America's Southwest. A company called Performance Aire, in Anaheim, California, specializes in Porsche air and, among other things installs a large vent outlet below the dash. Performance Aire also does conversions from CFC-based coolants to environmentally friendly HFC-134A coolants.

ABOVE
When you test the air conditioning as part of your own prepurchase inspection, this is where the thermometer goes. Factory specifications call for a minimum temperature of 32 degrees Fahrenheit.

Porsche ownership often serves as bookends to marriage and family life. The car leaves with the birth of children and reappears after they graduate college.

Yet child seats fit securely. The owners evaluated many seats in their search for one in which their young son could join them.

If the roof panel is there, the next major concern is the rear window. Earliest versions were plastic with zippers to remove the rear window, providing the benefit of a factory-installed roll bar in a fairly stiff open car. However, the zippers dry out and the plastic shrinks, splits, and discolors. Later models offered or provided a fixed glass window. They afforded better weather sealing, and by the mid-1980s, rear window defogging capability. However, this glass is a barometer indicating rear end damage. If the window has begun to crack, rear end body damage is likely. If the glass breaks, it can be extremely difficult and costly to replace. Repair shops admit to breaking another window or two as they try to install a replacement.

As with Cabrios, there are particular areas to inspect carefully. The Targa was introduced in 1967 and has changed over the decades. Please see your individual year for important information.

SECTION 9. ELECTRONICS

This applies mostly to 1989 1/2 model year cars and after. These often have aftermarket installed theft alarms, cellular telephones, radar detectors, and other add-ons. Improperly installed extras can rapidly run down batteries. Cutting into wires, unless you know exactly what you are doing, is a recipe for disaster. Not all hot leads are the same. Ask if the seller knows who did the work. If it was not a Porsche dealer, you may consider having your local dealer remove and reinstall these items. This becomes more important with each newer model year. Incorrectly wired accessories will drain batteries in 10 days, which can adversely affect onboard computers.

SECTION 10. WHAT, YOU SAY? EVERYTHING WORKS PERFECTLY?

And, secretly, it's the color combination you've wanted all your life? Now it's time to "do the deal."

10A. Pricing is a tough issue. Dealers, resellers, and independent service facilities all agree that you should buy the best car you can afford. Porsche resale values remain high; "blue book" type values are an estimate at best. Consider spending a few thousand dollars less than your maximum budget because you'll likely put that into the car almost immediately for major or minor services, fresh rolling stock, or dealing with cosmetic issues.

10B. Know what you can spend and do have all your financing in order so you can move quickly if, after the inspection, you have found the right car.

10C. Don't take a seller's word that something will be fixed, or that the problem really is nothing at all. If the problem is to be fixed, get that in writing and add in writing that you will conclude the deal only if the "problem is fixed to your satisfaction, pending another complete test drive."

This 1972 911S reveals the telltale proof of its age: the outside oil-filler cap cover on the passenger side. The idea for placing it there was to optimize weight balance; it lasted a single year and safety concerns relocated it to its original rear fender position.

The Turbo returned to Porsche's 911 lineup in September 1990 on the new Typ 964 platform. Engines developed 320hp and Porsche assembled these through 1992, producing something like 3,660 examples before the 3.6-liter replaced it.

10D. Living with the car: Everyone consulted for this book also suggests you set aside a monthly "allowance" to cover routine and not-so-routine expenses. If you save $250 a month, then when the annual minor and major service bills arrive for $2,500 to $3,000, you won't feel it.

CONCLUSIONS

If this entire chapter seems like an outline of don'ts, that's because you can figure out the do's.

There are two important do's.

Do thank the seller and go home if the engine is rough and balky, or if you find a car with wavy roof lines above or alongside the windshield or rear window, or along the doorsills and jambs, indicating serious crash damage.

Do be patient, however, because most of the cars out there are good ones.

Chapter 2
Open and Shut Cases:
Targas and Cabriolets

THE TARGA

Since Ferdinand Porsche's first car, 356-001, designed by Erwin Komenda and completed in May 1948, open air driving has been a priority with the company. Soon after the commercial introduction of his first Coupes, built in Gmünd, Austria, Porsche contacted Beutler Carrosserie in Thun, Switzerland, to produce open versions. When Beutler chose not to go further than the six the firm completed, Porsche worked with German coachbuilder Reutter in Stuttgart, starting in 1950.

It took roughly the same time span following the production introduction of the new 911 model, two years, before Porsche had an open car on that chassis. They named it the Targa after Porsche's numerous victories in the Targa Florio open road race in Sicily, but this car was not a full convertible. The factory did develop two prototype convertibles during 1965 and 1966, but Porsche feared the growing legislative might of U.S. government agencies. He worried that a full convertible might become illegal overnight. So Ferdinand Porsche's grandson, Butzi, who had designed the 911, created the roll bar–equipped Targa model. This design served several functions.

Porsche built its 911 models without a separate frame, using unitized chassis construction in which the car body serves as part of the car structure. Removing a roof is a serious consideration. The roll bar served to restore some—but not all—of the stiffness of the Coupes to this open car body. Butzi and the engineers augmented the floor pan with additional plates that wrapped up into the front passenger foot well and rear passenger seating area, as well as a stiffening bar just under the dashboard. Those who drive a number of Coupe and Targa (or later Cabriolet) models will notice the difference. It is slight, but it is perceivable.

If you are looking at buying one of the Targa models, one important fact to know is that there are no replacement roofs available any longer from the factory. Porsche did not produce extras, so if the car you are examining has a badly damaged or missing roof, this is a serious problem. But it is not insurmountable. Some Targa buyers acquire the car to convert to a full Cabriolet, and a number of specialist Porsche shops have acquired these "spares."

The tops disappear in several ways. If you must hurriedly reattach the top and the car is then towed backward, the top will blow off and probably be run over by traffic behind the tow truck, as the tow driver, not aware he should be watching for this, hurries your car to the service shop. The tops are not commonly stolen, but . . .

If the condition of the rest of the car is rough, then you can thank the seller and leave. If the roof panel is there but in rough condition, that is much less worry. They can be reconditioned or recovered, or the solid panels can be repainted.

The next major concern is the rear window. Early versions—from introduction in 1967 through optional availability as late as 1972—were plastic with zippers to remove the rear window. However, the zippers dry out and the plastic will shrink, split, or discolor. Later models provided a fixed glass window, first available in 1969. This is both good and bad. They provided better weather sealing, and by the mid-1980s they also included rear window defogging capability. However, this glass is a barometer that can indicate bigger problems. If the window has begun to crack, rear end body damage is likely.

Early models of the Targa removable top panels fold for storage. Make certain that they will unfold and lie flat in order to seal the roof. As with foldable roofs and convertibles, there are leaks. Check sun visors and interior panels for discoloration that may be water leaks, especially in older cars.

On models beginning in 1996, the Targa became a sliding roof model with an entire glass panel retracting. This is a very complicated system. For this unitized chassis, Porsche was able to attain only 75 percent of the stiffness of a regular Coupe. While this doesn't qualify the car as a "flexible flyer," it suggests a warning to buyers.

The owner's manual advises users to open or close the glass roof panel only on a level flat surface. This is important. Parking on a side incline or with one wheel in a pothole or another up a curb puts a twisting load on the unitized car body structure that will have its strongest effect at its weakest link, in this case, where there is movable glass, not where there is welded steel.

While this original-owner cabriolet is mostly stock, the owner fitted it with 17-inch Carrera Cup wheels with a deeper offset. This improved handling (and many would suggest, the car's appearance) without compromising safety. Many aftermarket wheels are stylish but do not possess the strength of Porsche's forged products.

Another "acquired" taste, this 1994 Speedster, based on the Typ 964 body style, appeals to as many enthusiasts as it repels. Porsche produced 469 for U.S. markets. This beautiful car is an excellent reason to insist on prepurchase inspections. This car was rebuilt after a serious accident and its owner, a talented mechanic and car painter, acquired it on a salvage title. Some less honest sellers may try to hide that fact.

The 1996 models arrived in the United States preceded by 15 service notices to dealers, describing pre-delivery work that the factory had determined needed to be done before car delivery to customers. In all, the 1996 and 1997 Targas were subject to something like 22 notices.

Porsche completely revised the roof glass transport system and mounted additional stiffening panels across the floor pan for 1998 models. So far, there have been no service notices or problems with that year's models.

There are so few of these glass-roof Targa models out there that one can only surmise that weather and wind sealing would remain a consideration and the operating motors and guiding tracks should be carefully inspected during routine dealer service.

Attraction to the Targa models is highly personal. The Targa is perhaps the most strongly defended or actively disliked of all Porsche 911 models. With the panel removed, it offers the benefits of a rollover-protection-equipped convertible while providing much more interior space and better security than either Porsche's 911 Cabrios or the Speedsters. However, its detractors point out that it is neither a true open car nor closed one. Because of this schizophrenic identity crisis, it is one of the least likely Porsches to be stolen and ranks alongside the 928 in that regard. This is a benefit you will realize when you insure your Targa.

THE CABRIOLETS AND SPEEDSTERS

Porsche introduced its Cabriolet in 1983 with a manual top and zippered glass window. Porsche made an electric lift standard equipment for model year 1987. If you are inspecting a Porsche Cabriolet or 911 Speedster, ask the seller to operate the top through at least two full open-and-close cycles. Do this only on a flat, level surface. Make certain that every function of the operation is working and that windows raise and lower as they are meant to do on later model cars with power roofs.

If one side appears out of sync with the other, if the top appears to be raising or lowering in a lopsided manner, stop immediately. The electrically operated tops are powered by cables similar to speedometer cables and they can "twist up," or bind up unequally, especially if the car is on a hillside where more weight is on the downhill side. Forcing a top at this point, especially closed, can break the bows, tearing the roof fabric. By the time all the damage is done, repair costs can reach several thousand dollars. This is why you ask the seller to do this test.

Do not stop the cycle with the top partly raised or lowered. Porsche uses four motors to operate the top and it actually relies on momentum at one point in the process to carry the top from one set of motors over to the other. If you stop the top at the wrong point—and no one seems to know where the right position is—you risk jamming or twisting the top.

This was one of Porsche's cleverest ideas, incorporating a large retracting glass roof within a coupe body. It enabled buyers to adapt to year-round weather conditions. Unfortunately, the first versions, in 1996 and 1997, were less rigid and some of the glass panels have been known to jam, necessitating a very expensive repair. For 1998, Porsche drastically reinforced and redesigned the system, providing much greater reliability.

With the roof shut, the car is nearly as quiet as a coupe. With the roof slid open, it offers much more protection from side and backwinds than a traditional cabriolet. Porsche introduced the glass-roof model in 1996. Operate the roof only on a flat, level surface with 1996 or 1997 models, as they can jam, requiring a very costly repair. Porsche redesigned the 1998 models and stiffened the car considerably, and no problems have been reported.

When the top is open, look for worn seals that meet the windshield. These let in wind noise and moisture. When the top is raised and secured, examine the liner for dirt, tears, and wear. Look carefully at the condition of the rear window.

On the second raise-and-lower cycle, look carefully into the roof well for leaves, sticks, or other debris. Do not reach in to retrieve them. Bows and struts cross each other during the opening and closing process and the risk of pinching or crushing a hand or finger is very great. Look only. Notice also the bows and struts that support the roof. Again, do not reach in.

Check to see that the roof fabric and liner have not torn away from the cross bows and that each joint is lubricated but not too heavily or recently (unless this is typical of the overall condition of a car that is very thoroughly maintained).

Look also for water stains that indicate either that the car got caught in the rain with its top down or that the top leaks around the windshield or the rear seals. Porsche prides itself on the quality of its convertible tops. (The 1983 Cabriolet photographed in this book is a case in point. Still driven by the original owner's son, the car also still has the original top and back plastic window. It is evidence of great care, and it's proof that great care pays off.)

The initial workmanship is extremely high and the mechanism that does the work, as already explained, is complex and clever. Repairs are complex and costly, and replacing a roof or rear window can dampen your enthusiasm for your new car.

With the 1989 and 1994 Speedster models, these roofs are manually operated and require that you follow a sequence of actions to release the roof, release and relocate the hard tonneau cover, stow the roof, and reset the cover. Because these are manually operated, you can visually and tactually inspect roof wells, bows, struts, and cloth. Be careful because the risk of pinching is still great.

If you purchase one of these models, ask that the seller demonstrate the raising and lowering operation for you several times. It is not difficult, though one source for this book referred to either operation as "the most complicated 30 minutes you'll ever spend with your car everyday." In truth, you can get the procedure down to 30 seconds, as was the case with the owners of both the 1989 and 1994 models photographed here. You may want to videotape the steps as the seller demonstrates.

Surprisingly, Cabrios are nearly as strong an emotional issue as are the Targas. Some Porsche enthusiasts wouldn't have one as a gift, fearing unitized body weakness as compared to the rock-rigid Coupes. Another source for this book disdainfully dismissed Cabrio purchasers as those who "just putt," as in, putt around, getting in the way of the more serious Coupe drivers.

Whatever model inspires you, remember the most important advice. There are lots of cars out there. If the one you're looking at is not right, thank the seller and leave.

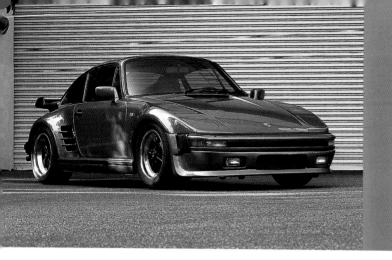

Chapter 3
Even with a Gray Market, There's Nothing Black and White

From 1975 until the late 1980s, a number of automobile conversion businesses flourished in the United States, providing to U.S. buyers cars they could not get through dealers. The car that drove this market, that kindled the desire, was the Porsche Turbo, introduced in 1973 and offered to European customers in late 1975. From 1976 until 1979, the European models, unrestricted by U.S. emissions and safety standards, developed more horsepower and were, for some U.S. buyers, more desirable. From 1980 until 1986, the period of time when Porsche chose not to offer its Turbos to U.S. customers, these converters flourished. Their solvency continued even after Porsche restored U.S. sales, because these outside vendors often got cars faster than dealer waiting lists could fill anxious desires.

Because the Environmental Protection Agency and the U.S. Department of Transportation emissions and safety standards were strict and their rules clearly published in black and white, standards of compliance were rigid and precise. However, because of human nature and the vast amounts of money involved, enforcement and ethics got a little murky. What was black and white became gray.

The gray-market importers and "federalizers" capitalized on an EPA-DOT loophole that allowed U.S. citizens a one-time importation, for personal use only, not greater than 2,500 miles per year, of a noncompliant vehicle. Within six months of the date the vehicle passed through U.S. Customs, the owner had to present it to one of several EPA-DOT testing centers in California, Texas, or New Jersey, for a "smog test" and a visual inspection demonstrating that front, rear, and side impact safety regulations had been complied with.

Customers generally paid the importer/federalizer to acquire the car in Europe and then gave them an extra $7,500 to $15,000 to "federalize" the car so it would comply and pass its tests. That is where the human nature element entered the equation.

Converting European front and rear bumpers, installing the pollution control machinery, and making the engine idle smoothly and run well were difficult challenges, even for major manufacturers to meet. In some cases, the independents didn't bother, or

merely short-cut their way through. The DOT did not require private owners to demonstrate 5-mile-per-hour bumpers by crashing the car. As a result, many cars gained 5-mile-per-hour bumpers that were welded to rigid frame members rather than bolted to energy-absorbing shock mounts. Side impact beams, meant to block another vehicle's caving in the door in a T-bone-type crash, were another area of creative compliance. Scott Hendry, owner of Scott's Independent, Inc., in Anaheim, California, has seen one Turbo in which both door side impact beams, required to be welded-in steel rod, were screwed-in, black-painted wooden broomsticks.

And therein lies the problem with the gray-market cars. You cannot possibly know what you are getting. And even if you got paperwork with the car telling you who performed the federalization, many of these companies disappeared in the late 1980s as the marketplace dissolved. Worse, while some of them were very responsible, capable mechanics and engineers, others, well . . . some of the others were not.

If this all reads like an across-the-board condemnation of gray-market cars, even that's a gray matter. In California and other states, emissions standards are subject to legislative action, and laws do change. What was required in 1976 or 1982 may not be needed today. Only a careful study of motor vehicle requirements in your state will tell you that.

Stepping out onto a limb here, generally, cars acquired new by the federalizers, purchased as new cars from cooperative dealers or factories, that have spent their entire life in the United States, are less of a risk. They still may have broomsticks for side impact beams, but a number of sources contacted for this book agreed that this brand of gray-market vehicle is a lesser risk. The car photographed for this chapter, this 1985 Slant Nose Turbo, was one such car. It was brought in new with nine others to a federalizer located in Scottsdale, Arizona.

Cars acquired used, brought into the United States with some miles on them, are, as everyone agreed, a huge risk. But again, how do you know, especially if the seller does know the car's history and has chosen to hide some information.

This is another reason to take any 911, even one that appears perfect such as this, to a competent mechanic for a prepurchase inspection. This otherwise complete-looking car is missing catalytic converters and other pollution-control equipment that could make it difficult to pass smog tests and license the car in many states.

This is the VIN number—WPOZZZ—and placement—at the lower left of the windshield just ahead of the steering wheel—to look for when identifying a gray-market car. It is hard to resist cars such as this one, but state laws make licensing them difficult, if not impossible. Insurance companies decline to cover them, and lending institutions will not finance them. The owner plans to use this for club racing events in the Midwest.

To identify the imported cars, look at the 17-digit VIN number. Those models destined for U.S. delivery will read WPOAA with 12 numbers and letters after the double A. If you see WPOZZZ, this is a non-U.S. model. On this model, it would be essential to have a prepurchase inspection done. People such as Scott Hendry, other highly experienced independents, and your local authorized

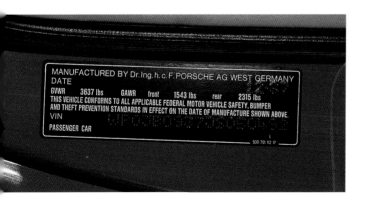

This is a 1988 Turbo VIN number for a U.S.-legal model. By 1988, the Turbo was legal again in the United States and gray-market federalizers had begun to disappear. That is another problem now for those with problems or complaints. Buyers of these cars have no recourse and are at legal and financial risk.

Porsche dealer can tell you what you're looking at and warn you better of your risks.

It is important to understand, however, that you cannot get financing to purchase this car. What's more, you may not be able to insure your gray-market car these days. If you're looking only for a "track" car, for weekend outings with Porsche Owners Club members on closed tracks, this would be an option for you. But for on-road use, you may find it is more trouble than it's worth.

The gray market imported not only Porsches but also a number of high-end Mercedes-Benz and BMW sedans and Ferrari 365 Boxer models. But the gray market for Porsche picked up steam particularly after England's Royal Auto Club officiated at a "supercar" test in June 1984 and determined the Turbo was "The Fastest Accelerating Production Car in the World," mastering the standing-start kilometer (.625 miles) at 135 miles per hour.

Check with your financial lender and with your insurance agent before you commit to purchasing one of these cars. In some states such as California, successful completion of a smog test is required before vehicle transfer can take place. If the seller offers to take care of this for you, with you not being present for the testing, beware. You may not be able to pull off the same miracle when you are required to retest it for license plate renewal a year or two later.

It took Porsche nearly eight years to settle on the design, engineering, and appearance of the 911 to replace its long-lived Typ 356. Part of that time was defining the new car. Where 356s had been 2+2 seaters, body designer Erwin Komenda urged Ferry Porsche to graduate to a true four-seater. Designers and modelers working for Ferry's son Butzi Porsche stuck with the 2+2 configuration. Meanwhile, Ferry's engine designers worked toward his goal of a 130-horspower engine for this new car. They fitted fuel injection to a four-cylinder Typ 616 engine, the 1.6-liter powerplant in the then-current 356C. But this didn't yield enough power. By 1961, engineers had begun designing an all-new pushrod-activated overhead-valve six cylinder. It proved to be too bulky for the "packages" Butzi's team were designing, so engineering chief Klaus von Rucker redesigned the valve train. By early 1962, the engine produced 120 horsepower DIN and when engineers increased displacement from 2.0 to 2.2-liter, they finally got 130, but this left little room for displacement increases for racing applications. Months later von Rucker left and Hans Tomala took over. This gave Ferry Porsche an opportunity to review his ongoing projects and he made a fateful change.

"Why should we make such a four-seater car when all the rest can do it better than we can?" he answered rhetorically when interviewed about the new car years later. He approved reducing the wheelbase from 94.5 to 87.0 inches, just 4.4 inches longer than production 356s. Butzi's designers slimmed the roof and side window lines, inclined the door's rear B-pillar forward, and made the windshield angle a bit steeper than earlier efforts, all to improve entry and interior room.

By early 1963, the many changes to the car led to a new Typ designation: 901. Engine designer Hans Mezger and engineer Ferdinand Piëch (Ferry's nephew) moved the camshafts to the outside of each of the cylinder heads in a single casting. Testing proved they'd reached their goal of 130 DIN horsepower, 148 SAE at 6,100 rpm.

As the engineers finished the car for manufacturing, they added ZF-produced rack-and-pinion steering adopted from their Typ 804 Formula One race car. They mounted the steering box midway between the front wheels to facilitate assembling left- or right-hand-drive models. This gave the steering wheel shaft a pair of universal joints and kinks that prevented the steering wheel from impaling the driver's chest in a head-on collision.

While they were developing the 901, Porsche engineers worked vigorously to launch a new Typ 904 racing coupe, which Butzi's designers also had created. The company's deadline to manufacture 100 of the Typ 904 cars to meet racing rules delayed 901 preproduction to May 1964.

Porsche started regular series production in September 1964 for model year 1965. Then French automaker Peugeot informed Porsche of a French trademark ruling giving it rights to production car numbering systems with a zero in the middle. During a two-week halt in assembly, Ferry redesignated the model the 911.

Most automotive journalists liked the car and praised the performance Porsche had derived from 2.0 liters. US magazines averaged eight seconds to 60 mph and found top speeds of 130 mph to be astonishing, something found only in cars with more cylinders costing much more money. However, several commented on the car's price, much higher than the still-in-production 356SC and more even than the recent 130hp Carrera 2 models.

Still, as with any first-year model of any automobile, the 1965 Porsche 911 had its faults. Without doubt these have been remedied years before now—by owners and certainly by Porsche in subsequent design improvements. They are mentioned here as a review, not a warning. Other issues, such as rust, remain a problem for the life of the car. Porsche's early undercoating held moisture against the steel. Oil leaks were a consistent problem.

These first year models sell for prices higher than those manufactured three or five years—or 30 or even 50 years—later because they are the first. A very serious coterie of enthusiasts recognize the technological accomplishment this car represented, no matter the price.

1965 Specifications "O" Series

Body Designation:		911 coupe
Price:		$5,990 POE New Jersey
Engine Displacement and Type:		Typ 901—1,991 cc (121.5 cid) SOHC
Maximum Horsepower @ rpm:		148 SAE gross @ 6,100 rpm
Maximum Torque @ rpm:		140 ft-lb @ 4,200 rpm
Weight:		2,360 pounds
0–62 mph:		9.0 seconds (*Road & Track*)
Maximum Speed:		132 mph (*Road & Track*)
Brakes:		ATE-Dunlop four-piston caliper disc brakes
Steering:		ZF rack-and-pinion
Suspension:	Front:	MacPherson struts, telescoping shock absorbers, lower wishbone, longitudinal torsion bars, antisway bar
	Rear:	telescoping shock absorbers, semi-trailing arms, transverse torsion bars
Tires:	Front:	165HR15
	Rear:	165HR15
Tire air pressure:	Front:	28 psi; Rear: 33 psi.
Transmission(s):		Typ 901/2, five-speed
Wheels:	Front:	4.5Jx15 stamped steel
	Rear:	4.5Jx15 stamped steel

What they said at the time—Porsche for 1965

Sports Car Graphic, January 1965

"Probably the biggest attraction in owning a Porsche is the mystique. All the engineers and professional men who are Gran Turismo buffs seem to wind up owning a Porsche. It's kind of a club, an in-group of people who think they're special because they've had the good taste to buy something special. At almost $6,500, the 911 is more than a Corvette, less than a Ferrari, and within not too many dollars of being a mighty expensive car. But to its owners, it's more than worth it."

Parts List for 1965 911s

These are items most commonly replaced during regular maintenance and routine daily operation. Prices quoted are for new factory parts at list price, not including installation labor. NLA means factory parts no longer available, so prices quoted are from aftermarket suppliers.

Engine:

1. Oil filter.................... $16.50
2. Alternator belt.......... $16.00
3. Starter...................... $433.15
4. Alternator (NLA) $150.00
5. Muffler $1,126.98
6. Clutch disc $531.42

Body:

7. Front bumper............. $1,991.98
8. Left front fender $2,053.12
9. Right rear quarter panel............. $1,959.18
10. Front deck lid $1,182.54
11. Front deck lid struts.. $43.22 each
12. Rear deck lid struts... $25.25 each
13. Porsche badge, front deck lid............. $201.18
14. Taillight assembly $854.45
15. Windshield (NLA)....... $509.00
16. Windshield weather stripping.................... $155.44

Interior:

17. Dashboard (NLA)....... $1,298.36
18. Shift knob $217.14
19. Interior carpet, complete (NLA) $1,800.00

Chassis:

20. Front rotor................. $205.80
21. Brake pads, front set $95.09
22. Koni rear shock absorber.......... $374.72
23. Front wheel (NLA)...... $50–$100
24. Rear wheel (NLA) ...... $50–$100

Ratings

1965 model	911 coupe only
Acceleration:	3
Comfort:	3
Handling:	2
Parts Availability:	2
Reliability:	1.5a

a - Replacing original Solex carburetors improves reliability; see text.

1965 Garage Watch
Problems with (and improvements to) Porsche 911 models.

1964 is considered as prototype/preproduction year. 1965 is first "regular production" model year. *NOTE: When Porsche introduced the 911, it told Porsche Club of America members they should not mount their PCA badges on the engine deck lid grille because it would disrupt airflow to the engine. "This is a high performance automobile and it needs all the air it can get."*

Factory original alternator produced only 35 amps, barely powerful enough for the time and too small for modern additions such as radar detectors, cell phones, etc.

Windshield and rear windows leak in corners when rubber seals get old and brittle.

Tensioners for the chain driving the overhead camshaft are a problem. Because of Porsche's sand-cast blocks on these early engines, it is not possible to replace failed tensioners with Carrera engine pieces. So most owners have gone to 930-engine-type tensioners or mechanical tensioners. This will already have been done. Examining owner's receipts will show when and how this was replaced.

Modern day gasoline has so many additives that it won't burn fully with the lower temperature-range spark plugs originally specified. Buyers/owners should run plugs at least two to four steps hotter. Car needs add-on capacitive discharge ignition (CDI) system to run well on today's gasoline.

Solex carburetors are hard to set up because the metal is very soft; very few are left because they simply wore out; most were converted to Webers.

Front parking lights and rear taillights are one-piece lens and reflector assembly. The rear taillights especially are seriously expensive but necessary for a "concours" restoration. Otherwise, reproduction lenses are available.

Rust in front shock towers and rear torsion bar housing areas, primarily on Midwest, Southeast, and East Coast cars but also California cars around San Francisco Bay Area.

Door handles malfunction because of problems with the door striker plate on the car body and the latch mechanism.

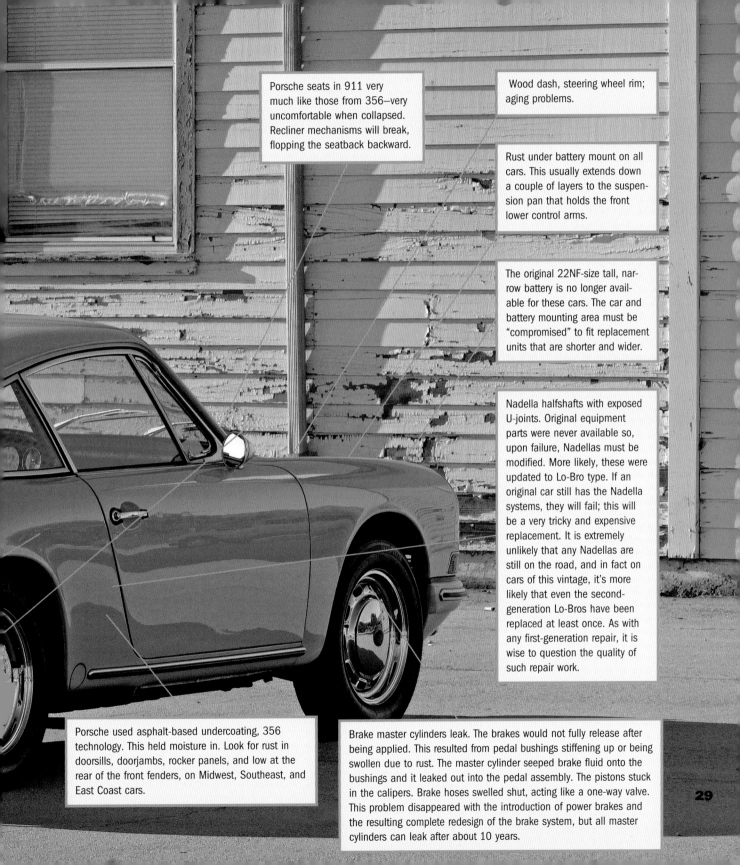

Porsche seats in 911 very much like those from 356—very uncomfortable when collapsed. Recliner mechanisms will break, flopping the seatback backward.

Wood dash, steering wheel rim; aging problems.

Rust under battery mount on all cars. This usually extends down a couple of layers to the suspension pan that holds the front lower control arms.

The original 22NF-size tall, narrow battery is no longer available for these cars. The car and battery mounting area must be "compromised" to fit replacement units that are shorter and wider.

Nadella halfshafts with exposed U-joints. Original equipment parts were never available so, upon failure, Nadellas must be modified. More likely, these were updated to Lo-Bro type. If an original car still has the Nadella systems, they will fail; this will be a very tricky and expensive replacement. It is extremely unlikely that any Nadellas are still on the road, and in fact on cars of this vintage, it's more likely that even the second-generation Lo-Bros have been replaced at least once. As with any first-generation repair, it is wise to question the quality of such repair work.

Porsche used asphalt-based undercoating, 356 technology. This held moisture in. Look for rust in doorsills, doorjambs, rocker panels, and low at the rear of the front fenders, on Midwest, Southeast, and East Coast cars.

Brake master cylinders leak. The brakes would not fully release after being applied. This resulted from pedal bushings stiffening up or being swollen due to rust. The master cylinder seeped brake fluid onto the bushings and it leaked out into the pedal assembly. The pistons stuck in the calipers. Brake hoses swelled shut, acting like a one-way valve. This problem disappeared with the introduction of power brakes and the resulting complete redesign of the brake system, but all master cylinders can leak after about 10 years.

In all, Porsche produced 3,389 of the 911 models in model year 1965. Yet, according to Tobias Eichele's book, *Porsche 911 Forever Young*, the factory manufactured 3,724 in 1966, barely 550 more than 1965, and explaining the difficulty you will have in finding either one to look at or purchase.

Porsche did very little to change the car from model year 1965 to 1966 production, and only by carrying a serial number guide along with you could you easily differentiate between the first and second year.

It was, however, a 1966 model that longtime Porsche enthusiast and legendary automotive journalist Dennis Jenkinson got to enjoy for "a glorious week." During that period, he put 1,200 miles on the car, indicating what Jenks considers "glorious," and also how easy the car was to drive that far at that time.

"Its first-class stability, hard-working engine, perfect gearbox, accurate steering, hard cornering propensities, and one-piece feel make it a real GT car, a car that is intended for hard motoring. The harder you drive it the more it seems to come alive, and you can almost hear it chuckling to itself as you really begin to use it the way Dr. Porsche meant it to be used."

The 911, Jenks continued, "set standards toward which everyone should endeavor to arrive."

Engineers expended little effort on the second year model. It's understandable. They had their plates full with an aggressive racing program of new six-cylinder Typ 906 models. They also had a four-cylinder production model, the Typ 912, using the 911 body, ready to introduce in April 1965. This car used the 1600 SC engine from just-discontinued 356SC models. Porsche priced it at $1,412 less than its six-cylinder counterpart—$4,063 at the factory compared to $5,475—and 912s nearly outsold the 911 by two-to-one. Porsche produced 6,401 of the 912s, compared to 3,724 911s.

Because there are no differences between 1966 models and those from 1965, a prudent shopper would look for the same traits, performance, and problems. No new problems appeared, but none of the first ones were yet remedied. In particular, Porsche

fitted Nadella half-shafts with exposed U-joints. These, like many early engineering choices, have failed long ago and Porsche itself never stocked replacements. The company introduced Lo-Bro type joints in 1967 and any 1965 or 1966 model will have gone through one or more sets of these through routine wear. Another problem occurred with timing chain tensioners in the crankcases made of sand-cast aluminum. While the later Carrera-type tensioners will not work in the sand-cast blocks, most owners will have gone to 930 Turbo-type tensioners or a mechanical tensioner.

In addition, the early cars are susceptible to corrosion both in the body at the doorsills and jambs and around the windshield. Check also the areas down low at the rear of the front fenders. Most troubling is rust in the rear torsion bar area and under the battery. If this has penetrated down several layers (and it can) it reaches the suspension pan that holds the front lower control arms.

If you are seriously considering one of these models that predate the introduction of galvanizing (starting with model year 1976), you will want to look underneath the battery or examine repair receipts to make certain this work has been done. If you are going this far, check behind the headlights in the buckets as well. This area captures rain and the drain plug often fills from beneath with rain-splashed or tire-thrown road grime. Although true of all Porsches, a prepurchase inspection is especially recommended on pregalvanized car bodies built before 1976.

Now, decades later, it would be wise to expect that original oil and brake lines need replacement, especially those that run through the passenger compartment and are now aging from prolonged exposure to heat and ozone.

In mid-model year 1966 (production starting in February), Porsche replaced the Solex "overflow" type carburetors with Weber triple-throated down-draft models. The Solex carburetors were cast in very soft metal and as they have aged they are increasingly hard to keep in tune, requiring great experience and very good hearing. Many mechanics will recommend converting a Solex car to Webers for tuning ease and reliable running.

1966 Specifications "O" Series

Body Designation:		911 coupe
Price:		$5,990 POE New Jersey
Engine Displacement and Type:		Typ 901—1,991 cc (121.5 cid) SOHC
Maximum Horsepower @ rpm:		148 SAE gross @ 6,100 rpm
Maximum Torque @ rpm:		140 ft-lb @ 4,200 rpm
Weight:		2,360 pounds
0–62mph:		9.0 seconds (*Road & Track*)
Maximum Speed:		132 mph (*Road & Track*)
Brakes:		ATE Dunlop Four-piston caliper disc brakes
Steering:		ZF rack-and-pinion
Suspension:	Front:	MacPherson struts, telescoping shock absorbers, lower wishbone, longitudinal torsion bars, antisway bar
	Rear:	telescoping shock absorbers, semi-trailing arms, transverse torsion bars
Tires	Front:	165HR15
	Rear:	165HR15
Tire air pressure:	Front:	28 psi; Rear: 33 psi.
Transmission(s):		Typ 901/2, five speed
Wheels:	Front:	4.5Jx15 stamped steel
	Rear:	4.5Jx15 stamped steel

What they said at the time—Porsche for 1966

Car and Driver, April 1966

"No contest. This is the Porsche to end all Porsches—or rather, to start a whole new generation of Porsches. It's one of the best Gran Turismo cars in the world, certainly among the top three or four.

"Only yesterday, the 356 seemed ahead of its time. Today you realize its time has passed; the 356 leaves you utterly unimpressed and you can't keep your eyes off the 911. The 911 is a superior car in every respect . . . the stuff legends are made of."

Parts List for 1966 911s

These are items most commonly replaced during regular maintenance and routine daily operation. Prices quoted are for new factory parts at list price, not including installation labor. NLA means factory parts no longer available, so prices quoted are from aftermarket suppliers.

Engine:

1. Oil filter $16.50
2. Alternator belt $16.00
3. Starter $433.15
4. Alternator (NLA) $899.00
5. Muffler $1,126.98
6. Clutch disc $531.42

Body:

7. Front bumper $1,991.98
8. Left front fender $2,053.12
9. Right rear quarter panel $1,959.18
10. Front deck lid $1,852.54
11. Front deck lid struts .. $43.22 each
12. Rear deck lid struts ... $25.25 each
13. Porsche badge, front deck lid $201.18
14. Taillight assembly $854.45
15. Windshield (NLA) $509.00
16. Windshield weather stripping $155.44

Interior:

17. Dashboard (NLA) $1,298.36
18. Shift knob (5-speed) (NLA) $217.47
19. Interior carpet, complete (NLA) $1,800.00

Chassis:

20. Front rotor $205.80
21. Brake pads, front set $95.09
22. Koni rear shock absorber $374.72
23. Front wheel (NLA) ...... $50–$100
24. Rear wheel (NLA) ...... $50–$100

Ratings

1966 model	911 coupe only
Acceleration	3
Comfort	3
Handling	2
Parts	1
Reliability	1.5a

a - Replacing original Solex carburetors improves reliability; see text.

1966 Garage Watch
Problems with (and improvements to) Porsche 911 models.

The 1965 and 1966 cars are virtually identical. There were no new problems with Porsche factory cars or options. However, aftermarket suppliers began offering electric power sunroofs.

Rust under battery mount in all cars, extending to suspension pan holding lower front control arms, as from 1965.

The original 22NF-size tall, narrow battery is no longer available for these cars. The car and battery mounting area must be "compromised" to fit replacement units that are shorter and wider, as from 1965.

Wood dash, steering wheel rim; aging problems. Porsche also offered optional gauges and optional woods for the dash.

Rust in front shock and rear torsion bar mounts, non-California cars, as from 1965.

Nadella halfshafts with exposed U-joints, as from 1965, but only until mid-1966 model year. Then Porsche replaced them with Lo-Bro system with constant velocity joints weather-and-lubrication-sealed inside rubber boots. But these boots tear, allowing grease to leak out and dirt to get in. As the CVs start to fail, you can hear a knocking or loud clicking from the CV joint.

Brake master cylinders leak. The brakes would not fully release after being applied. This resulted from pedal bushings stiffening up, or being swollen due to rust. The master cylinder seeped brake fluid onto the bushings, and it leaked out into the pedal assembly. The pistons stuck in the calipers. Brake hoses swelled shut, acting like a one-way valve. This problem disappeared with the introduction of power brakes and the resulting complete redesign of the brake system, but all master cylinders can leak after about 10 years.

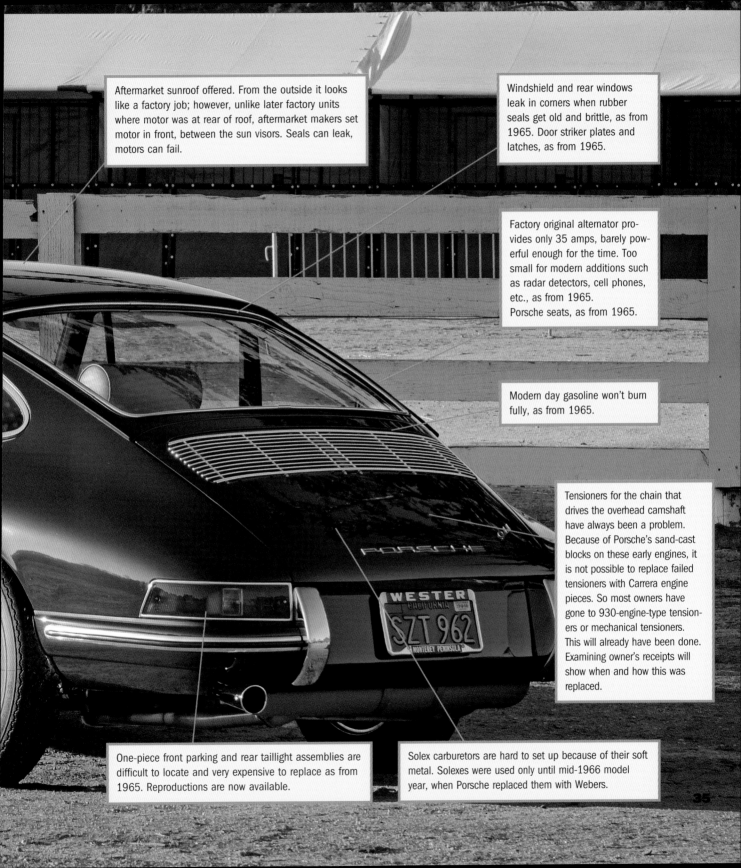

Aftermarket sunroof offered. From the outside it looks like a factory job; however, unlike later factory units where motor was at rear of roof, aftermarket makers set motor in front, between the sun visors. Seals can leak, motors can fail.

Windshield and rear windows leak in corners when rubber seals get old and brittle, as from 1965. Door striker plates and latches, as from 1965.

Factory original alternator provides only 35 amps, barely powerful enough for the time. Too small for modern additions such as radar detectors, cell phones, etc., as from 1965.
Porsche seats, as from 1965.

Modern day gasoline won't burn fully, as from 1965.

Tensioners for the chain that drives the overhead camshaft have always been a problem. Because of Porsche's sand-cast blocks on these early engines, it is not possible to replace failed tensioners with Carrera engine pieces. So most owners have gone to 930-engine-type tensioners or mechanical tensioners. This will already have been done. Examining owner's receipts will show when and how this was replaced.

One-piece front parking and rear taillight assemblies are difficult to locate and very expensive to replace as from 1965. Reproductions are now available.

Solex carburetors are hard to set up because of their soft metal. Solexes were used only until mid-1966 model year, when Porsche replaced them with Webers.

35

Two years after introducing the Typ 901 prototype at Frankfurt, Porsche unveiled its third variation, its "open" 911, in September 1965. It took the company another year after the unveiling to put this open car, similar to the 901, in production.

Butzi Porsche advocated a truly open car, but his father feared that the burgeoning safety regulations in the United States might outlaw open cars altogether. Further, there was the cost of developing a new, stronger chassis. The rigidity that a roof provided had to be replaced somehow. Designers Gerhard Schroder and Werner Trenkler, developed a roadster in which a rigid roll bar served as the soft roof's main support bow, disguising the roll bar in a brushed stainless steel sheath. They originally planned this bar as collapsible, but the cramped interior allowed no storage space.

They named the car the "Targa" to honor their own numerous victories at the race in Sicily. Following its introduction, it took the engineers another year to add reinforcement to the floor pan that wrapped up into the front footwells and the rear jumpseat area, and to make the car production-ready. Porsche's zippered, removable back window of plastic was not a perfect design, because if the air temperature was cooler than 60 degrees Fahrenheit, the plastic and the zipper shrunk and the window could not be refitted. And if it was in place, in cooler weather the seams stretched, allowing in moisture and cold air.

At the Frankfurt show in 1966, Porsche introduced a 911 model that filled out the line. Just as there had been "Normal," "Super," and "Carrera" models in the 356 line, the 912 replaced the normal, and the standard 911 remained as the middle rung of Porsche's performance ladder. On top now was the new "Super," the 911S.

Engineers replaced the Solex floatless system with downdraft Webers, and carried over that change to regular model 911s as well. These improvements, plus spark and valve timing changes, increased horsepower output to 160 DIN and 180 SAE for the S models. To visually differentiate the engines, Porsche changed the cooling fan fiberglass shroud from black on the 911 to red on the S versions, starting a pattern it would maintain for years.

Porsche introduced a matched five-speed transmission for this new S model, with a lower first gear and a higher top gear. Porsche fitted forged aluminum wheels made by Fuchs as standard equipment on the S. To improve handling, engineers added an antisway bar on the rear, the first time any production automobile came standard with one. It also provided ventilated disc brakes on these models.

Porsche had improved the car door locks for all three models, and it also improved the heat exchange bosses that supplied hot air to the cockpit heater. The company had a good year in 1967, manufacturing 6,645 of its 911 models (as well as 6,472 of the 912s).

As an aside, it's important to note a very limited production 911 model that appeared only in 1967, the legendary "R" model. The brainchild of brilliant engineer Ferdinand Piëch, it was created as a prototype to launch a broader production run, intended to qualify the 911 model for various international racing series. The 911R model interiors were spartan, stripped of carpets, insulation, and headliner materials. The factory formed thinner gauge steel for door skins, and front and rear deck lids, saving precious pounds wherever it could. Larger Weber 46-millimeter-diameter carburetors, twin spark plugs, and a higher engine speed provided 210 horsepower DIN (at 8,000 rpm). The accomplishments of these cars in races and endurance runs founded the racing heritage of the 911. These models seldom change hands, and on the rare occasions they do come up for sale, a phone call or two seals the transaction. Usually such cars command prices in the high sixfigure range. There are replicas in existence but these are known and accounted for.

The 1967 S models founded their own legends and a fine owner organization exists to document and preserve these cars through the 1973 model year. The Early S Registry is located in Newport Beach, California. If you own one of these or are searching for one, contact the registry at www.early911registry.org.

1967 Specifications "O" Series

Body Designation:		911 coupe, Targa, 911S
Price:		911 coupe: $5,990 POE New Jersey
		911 Targa: $6,490
		911S: $6,990
		911S Targa: $7,490
Engine Displacement and Type:		911: Typ 901/14 - 1,991 cc (121.5 cid) SOHC Weber 40IDS carburetors
		911S: Typ 901/02 - 1,991 cc SOHC
		Weber 40IDA carburetors
Maximum Horsepower @ rpm:		911: 148 SAE @ 6,100 rpm
		911S: 180 SAE @ 6,600 rpm
Maximum Torque @ rpm:		911: 140 ft-lb @ 4,200 rpm
		911S: 144 ft-lb @ 5,200 rpm
Weight:		911: 2,360 pounds
		Targa: 2,410 pounds
		911S: 2,279 pounds (factory)
0–60 mph:		911: 7.3 seconds (*Motor Trend*)
		Targa: 9.3 seconds
		911S: 6.9 seconds (*Motor Trend*)
Maximum Speed:		911: 138 mph (*Motor Trend*)
		911S: 152 mph (*Motor Trend*)
Brakes:		ATE-Dunlop Four-piston caliper disc brakes
Steering:		ZF rack-and-pinion
Suspension:	Front:	MacPherson struts, telescoping shock absorbers, lower wishbone,
		longitudinal torsion bars, antisway bar
	Rear:	telescoping shock absorbers, semi-trailing arms, transverse torsion bars; S adds antisway bar
Tires:	Front:	165HR15
	Rear:	165HR15
Tire air pressure:	Front:	28 psi; Rear: 33 psi
Transmission(s):		911 (U.S.) 902/0 4-speed in 1967 (optional 5-speed: $80)
		911S: Typ 901/2, 5-speed
Wheels:	Front:	4.5Jx15 stamped steel
		4.5Jx15 Fuchs on S, optional on 911
	Rear:	4.5Jx15 stamped steel
		4.5Jx15 Fuchs on S, optional on 911

What they said at the time—Porsche for 1967

Car and Driver, January 1967

"Oversteer is back—and Porsche's got it! Early Porsches had it too, and now it has come full circle. For a 2-liter sports car, a quarter-mile in 15.2 seconds at 92 miles per hour ranks with building a replica of the Great Pyramid of Cheops overnight. . .

"Each successive Porsche has been the ultimate Porsche, which is akin to its being the ultimate luxury GT car. The 911S surely must be the all-time high. Where can Porsche go from here? Build a car with disappearing headlights?"

Parts List for 1967 911s

These are items most commonly replaced during regular maintenance and routine daily operation. Prices quoted are for new factory parts at list price, not including installation labor. NLA means factory parts no longer available, so prices quoted are from aftermarket suppliers.

Engine:

1. Oil filter.................... $16.50
2. Alternator belt.......... $16.00
3. Starter...................... $433.15
4. Alternator (NLA) $899.00
5. Muffler $1,126.98
6. Clutch disc............... $531.42

Body:

7. Front bumper............. $1,991.98
8. Left front fender........ $,2053.12
9. Right rear quarter panel............. $1,959.18
10. Front deck lid............ $1,852.54
11. Front deck lid struts.. $43.22 each
12. Rear deck lid struts... $25.25 each
13. Porsche badge, front deck lid............. $201.18
14. Taillight assembly...... $854.45
15. Windshield (NLA)....... $509.00
16. Windshield weather stripping $155.44

Interior:

17. Dashboard (NLA)....... $1,298.36
18. Shift knob (5-speed) (NLA)........................ $217.47
19. Interior carpet, complete (NLA) $1,800.00

Chassis:

20. Front rotor................. $205.80
21. Brake pads, front set.................... $95.09
22. Koni rear shock absorber $374.72
23. Front wheel (NLA)...... $50–$100
24. Rear wheel (NLA) ...... $50–$100

Ratings

1967 models

	911 coupe	911S coupe	911 Targa	911S Targa	911R coupe
Acceleration	3	4	3	4	5
Comfort	3	3	3	3	2
Handling	2	3	2	2	4
Parts	2	3	1b	1b	1
Reliability	3	3	3	3	3

b - Targa roofs no longer available from Porsche.

1967 Garage Watch
Problems with (and improvements to) Porsche 911 models.

Many problems carried over for 1967, and the introduction of factory sunroofs and the Targa version introduced new concerns. This was the last year of instruments with green numbers on black backgrounds. Asphalt-based undercoating promotes rust, as from 1965. Rust in front shock towers and rear torsion bar housing area, as from 1965.

The original 22NF-size tall, narrow battery is no longer available for these cars. The car and battery mounting area must be "compromised" to fit replacement units that are shorter and wider, as from 1965.

Rust under battery mount down to suspension pan, all cars, as from 1965.

Aluminum inserts replaced wood dash in 911. Factory offered standard gauges for 911 (and 912) that included warning lights and optional gauges. 911S models retained their gauges, but on lesser 911 models, warning lights replaced the instruments.

First year of Targa with zip-out windows. All the roofs leak both wind noise and water. Weather seals dry out, crack. Finding replacements for early models is difficult, and all years are expensive. Two-piece fiberglass folds for storage. Plastic material cracks, crazes (a craze is a surface crack that, when abundant, gives the plastic a frosted appearance that poorly diffuses light and image transmission through the plastic), zippers jam or break. Factory Targa tops are no longer available; dealers are out of stock. However, independent shops often buy them from converters who cut off Targa roofs to convert cars to cabriolets.

Brake master cylinders leak. The brakes would not fully release after being applied. This resulted from pedal bushings stiffening up, or being swollen due to rust. The master cylinder seeped brake fluid onto the bushings, and it leaked out into the pedal assembly. The pistons stuck in the calipers. Brake hoses swelled shut, acting like a one-way valve. This problem disappeared with the introduction of power brakes and its resulting complete redesign of the brake system, but all master cylinders can leak after about 10 years, as from 1965.

Factory or aftermarket sunroofs (identifiable because aftermarket makers put motor between windshield visors). Seals can leak wind and rain.

911S introduced, offered only one year because of smog restrictions introduced for 1968. Car used leatherette dash inserts instead of wood. While wood dash inserts in 1965 and 1966 models often dried out and leatherette was introduced as improvement and upgrade, this material also cracked and is difficult and costly to locate and replace.

Windshield and rear windows leak in corners when rubber seals get old and brittle. Door striker plates and latches, as from 1965.

Modern day gasoline won't burn fully, as from 1965.

Chain tensioners for the chain driving the overhead camshaft have always been a problem. Because of Porsche's sand-cast blocks on these early engines, it is not possible to replace failed tensioners with Carrera engine pieces. So most owners have gone to 930 tensioners or mechanical tensioners. This will certainly have already been done. Examining owner's receipts will show when and how this was replaced.

Factory original alternator produced only 35 amps, barely powerful enough for the time. Too small for modern additions such as radar detectors, cell phones, etc., as from 1965. Porsche seats, as from 1965.

One-piece front parking and rear taillight assemblies hard to find and costly to replace, as from 65; reproductions available.

Dozens of little details marked the 1968 models as different from the first three years. Many of these came as a result of meeting vehicle safety and exhaust emissions regulations imposed on carmakers worldwide by a protective American government. For Porsche, these were necessary changes. Half its 911 production cars were sold to U.S. customers, 5,400 in 1967 alone.

To meet U.S. safety regulations, Porsche redesigned its door handles, eliminating the outside push button and replacing it with a lever inside the handle that one squeezed to open, enlarged 911 mirrors, introduced matte-black finished wiper blades to eliminate distracting glare and reversed their placement and sweep angle so they parked below the driver's eyes. American regulators demanded twin-circuit brake systems requiring tandem master cylinders and separate hydraulic brake fluid circuits for front and rear brakes, eliminating total brake loss in the event of a system failure or fluid leak.

Porsche fitted stronger windshields and seat belt mounts for the rear jump seats. It increased wheel rim width from 4.5 inches to 5.5 inches for the 15-inch tires.

But it was U.S. government exhaust emissions standards that had the greatest effect on enthusiasts. Porsche could not meet the problems of cleaning the exhaust of the powerful S engines in time for the 1968 model year. So it did not export these to the United States. Instead, it installed the standard 911 engine into the more fully equipped 911S bodies and labeled this model the 911L, so American customers had the 911L, the base 911, and the 912 for their three choices. In Europe, the S remained available, and Porsche, seeking to provide a lower-cost alternate, introduced the 911T, selling for even less than the base 911. Not quite as stripped out as the 1967 R model, the T still weighed about 77 pounds less than the base model, making it a desirable package for European racing series.

To meet the United States' first exhaust emission standards, Porsche engineer Hans Mezger and his crew installed air pumps. These worked like an air supercharger, driven by a belt pulley, forcing in extra outside air (without the additional fuel mixture of a real supercharger). But instead of going into the intake valves, these delivered warm air to the exhaust ports, close to the valve heads where this helped oxidize the exhaust hydrocarbons and carbon monoxide into water vapor and carbon dioxide vapor.

The engines did not like this system. The air pumps caused backfiring and the engines bucked at low rpm and balked at higher speeds. Owners found them hard to maintain and the engines needed tuning more often. Ironically, U.S. specification cars barely met the emissions standards with the air pumps and eventually the Environmental Protection Agency (EPA) recognized their limited value. The EPA eventually allowed 1968 912 owners to eliminate them altogether, and Porsche offered plugs to seal the engine cooling shrouds. Virtually every 911 owner followed suit even though, officially, the waiver did not extend to the six-cylinder engines.

Some Targa roofs leaked cold air and moisture through the roof panel seals and the zippered plastic back panel. American complaints got so loud that as an option the factory replaced the removable plastic window panel with a fixed glass back window that was air- and weather-tight.

Engineers developed a semiautomatic transmission, the Sportomatic, specifically aimed at attracting new American customers. The new transmission, the Typ 905, incorporated a torque converter with Porsche's existing four-speed transmission. What made it seem sports-car-like was the gearshift lever that, whenever the driver moved it, disengaged a clutch. The base of the lever tripped a microswitch that mechanically operated a vacuum servo-motor to disengage the clutch. The slightest touch on the lever disengaged the clutch.

Still, because both Corvette and Jaguar in the mid-1960s offered automatic as well as manual shift transmissions to an American marketplace quickly filling with drivers who hadn't mastered a clutch pedal, Porsche felt it had to compete.

Neither the T nor the S were legally imported to the United States for the 1968 model year. However, they are included in the technical specifications and parts prices because many of them have subsequently been imported to the United States. Their age precludes any need to comply with original smog regulations.

1968 Specifications "A" Series

Body Designation:		911T, 911L
Price:		911T: $6,150 POE New Jersey
		911T: $6,430 w/Sportomatic
		911L: $6,790
		911L: $7,070 w/Sportomatic
		911L: $7,190 w/Targa
Engine Displacement and Type:		Typ 901/14 – 1,991 cc (121.5 cid) SOHC Weber 46IDA 3C carburetors, with Exhaust Gas Recirculation (EGR) for U.S. 911 and 911L cars
Maximum Horsepower @ rpm:		148 SAE @ 6,100 rpm
Maximum Torque @ rpm:		145 ft-lb @ 4,200 rpm
Weight:		2,360 pounds T, L coupe
		2,420 pounds coupe w/Sportomatic
		2,410 pounds Targa
0–62 mph:		9.1 seconds (factory, coupe)
		9.3 seconds (factory, Targa)
		10.3 seconds (*Road & Track*, T w/Sportomatic)
Maximum Speed:		911T L: 131 mph (factory)
		Sporto: 117 mph (*Road & Track*)
Brakes:		ATE Dunlop Four-piston caliper disc brakes
Steering:		ZF rack-and-pinion
Suspension:	Front:	MacPherson struts, telescoping shock absorbers, lower wishbone, longitudinal torsion bars, antisway bar
	Rear:	telescoping shock absorbers, semitrailing arms, transverse torsion bars; L adds antisway bar
Tires:	Front:	165HR15, optional 185HR15
	Rear:	165HR15, optional 185HR15
Tire air pressure:	Front:	31 psi; Rear: 33 psi.
Transmission(s):		911T, L: 902/0 4-speed U.S.
		911T: 905/0 Sportomatic U.S.
		911L: 906/0 Sportomatic U.S.
Wheels:	Front:	5.5Jx15 Fuchs on T, L optional 6x15
	Rear:	5.5Jx15 Fuchs on T, L optional 7x15

What they said at the time–Porsche for 1968

Sports Car Graphic, March 1968

By Bob Kovacik

"Putting an automatic transmission in a Porsche is like artificial insemination: It's no fun anymore. At least that was our first impression.

"Bill Haworth, Porsche and Volkswagen public relations director in the West, told us they'd been waiting for this transmission for a long time. 'It's the type of thing that will bring us more customers from the professional ranks—doctors, lawyers, men and women, too, who don't want to be bothered with continual shifting in heavy traffic, but want a sports car.'"

Parts List for 1968 911s

These are items most commonly replaced during regular maintenance and routine daily operation. Prices quoted are for new factory parts at list price, not including installation labor. NLA means factory parts no longer available, so prices quoted are from aftermarket suppliers.

Engine:

1. Oil filter..................... $16.50
2. Alternator belt........... $16.00
3. Starter....................... $433.15
4. Alternator (NLA) $899.00
5. Muffler $1,126.98
6. Clutch disc $531.42

Body:

7. Front bumper............. $1,991.98
8. Left front fender........ $2,053.12
9. Right rear quarter panel $1,959.18
10. Front deck lid............ $1,852.54
11. Front deck lid struts.. $43.22 each
12. Rear deck lid struts... $25.25 each
13. Porsche badge, front deck lid............. $201.18
14. Taillight assembly...... $854.45
15. Windshield (NLA)....... $509.00
16. Windshield weather stripping $155.44

Interior:

17. Dashboard (NLA)....... $1,298.36
18. Shift knob (5-speed) (NLA)......................... $217.47
19. Interior carpet, complete (NLA) $1,800.00

Chassis:

20. Front rotor................. $205.80
21. Brake pads, front set.................... $95.06
22. Koni rear shock absorber $374.72
23. Front wheel (NLA)...... $50–$100
24. Rear wheel (NLA) ...... $50–$100

Ratings

1968 models, manual transmission

	911 coupe	911T coupe	911S coupe	911L coupe
Acceleration	3	2.5	4c	3
Comfort	3	2	3c	3
Handling	2.5	3	3.5c	3
Parts	2	2	3c	2
Reliability	2.5	3	3c	2.5

c – 911S 1968 model not originally sold in U.S. Some have been brought in.

1968 models, manual transmission, continued

	911 Targa	911T Targa	911S Targa	911L Targa
Acceleration	3	2.5	4c	3
Comfort	3	2	3c	3
Handling	2	2.5	3c	2.5
Parts	1b	1b	1bc	1b
Reliability	2.5	3	3c	2.5

b - Targa roof no longer available from Porsche.
c - 911S 1968 model not originally sold in U.S. Some have been brought in.

1968 models, Sportomatic transmission

	911 coupe	911T coupe	911S coupe	911L coupe
Acceleration	2.5	2	3.5c	2.5
Comfort	3	2	3c	3
Handling	2.5	3	3.5c	3
Parts	1d	1d	1d	1d
Reliability	1e	1e	1e	1e

c - 911S 1968 models not originally sold in U.S. Some have been brought in.
d - Sportomatic parts, gaskets extremely hard to find, costly to buy.
e - Sportomatic repair difficult, expensive, may fail quickly. See text.

Ratings continued

	911 Targa	911T Targa	911S Targa	911L Targa
Acceleration	2.5	2	3.5c	2.5
Comfort	3	2	3c	3
Handling	2	2.5	3c	2.5
Parts	1bd	1bd	1bd	1bd
Reliability	1e	1e	1ce	1e

b - Targa roofs no longer available from Porsche.

c - 911S 1968 models not originally sold in U.S. Some have been brought in.

d - Sportomatic parts, gaskets extremely hard to find, costly to buy.

e - Sportomatic repair difficult, expensive, may fail quickly. See text.

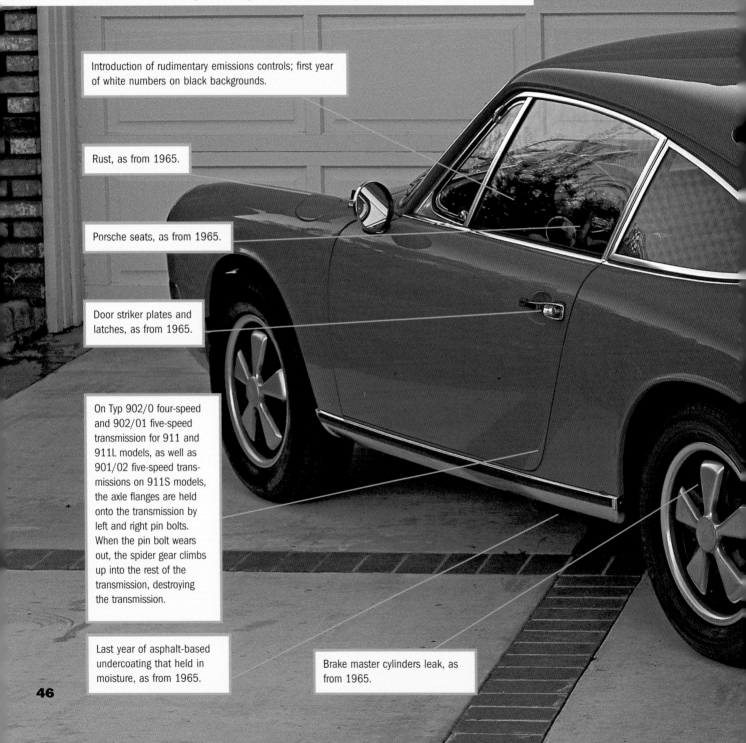

1968 Garage Watch
Problems with (and improvements to) Porsche 911 models.

Introduction of rudimentary emissions controls; first year of white numbers on black backgrounds.

Rust, as from 1965.

Porsche seats, as from 1965.

Door striker plates and latches, as from 1965.

On Typ 902/0 four-speed and 902/01 five-speed transmission for 911 and 911L models, as well as 901/02 five-speed transmissions on 911S models, the axle flanges are held onto the transmission by left and right pin bolts. When the pin bolt wears out, the spider gear climbs up into the rest of the transmission, destroying the transmission.

Last year of asphalt-based undercoating that held in moisture, as from 1965.

Brake master cylinders leak, as from 1965.

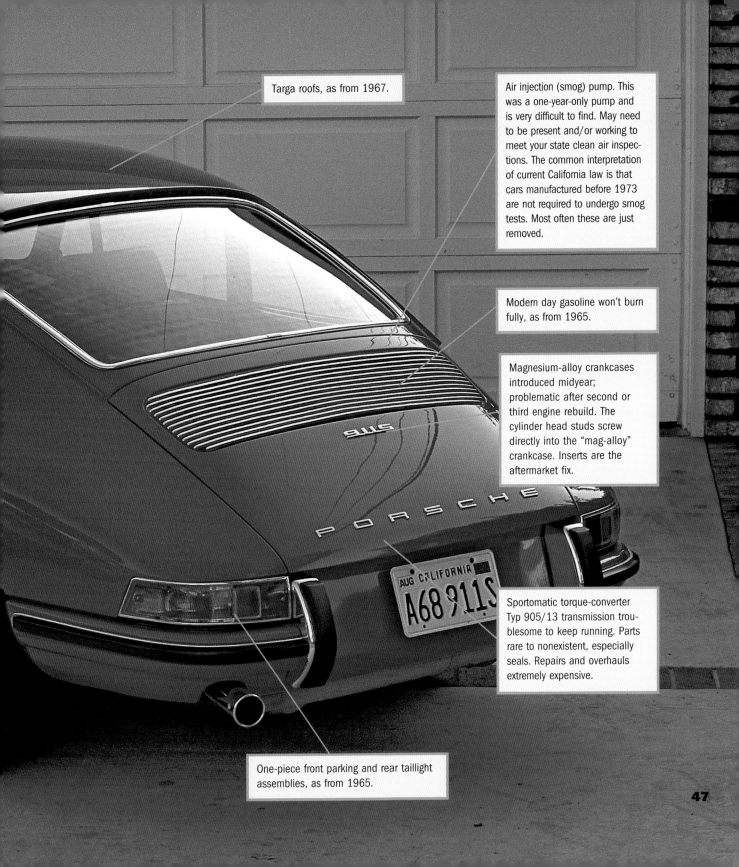

Targa roofs, as from 1967.

Air injection (smog) pump. This was a one-year-only pump and is very difficult to find. May need to be present and/or working to meet your state clean air inspections. The common interpretation of current California law is that cars manufactured before 1973 are not required to undergo smog tests. Most often these are just removed.

Modern day gasoline won't burn fully, as from 1965.

Magnesium-alloy crankcases introduced midyear; problematic after second or third engine rebuild. The cylinder head studs screw directly into the "mag-alloy" crankcase. Inserts are the aftermarket fix.

Sportomatic torque-converter Typ 905/13 transmission troublesome to keep running. Parts rare to nonexistent, especially seals. Repairs and overhauls extremely expensive.

One-piece front parking and rear taillight assemblies, as from 1965.

1969

Chapter 8
The "B" Series—
Big Changes Broaden
Horizons, and Bosch
Injects the Fuel

Revised carburetion and intake and exhaust manifold design eliminated the need for air pumps on U.S. models in 1969. In addition, Porsche brought fuel injection technology to the street on 911S models and on a new midlevel worldwide model, the 911E.

This injection system drove the fuel pump by a cogged belt off the end of one of the overhead camshafts. Each cylinder had a pump plunger that squirted fuel into an inlet port. It was reliable at very high engine speeds and over long periods of running. And it offered the additional benefit, for emissions-controlled engines, of allowing very precise adjustment of the fuel-air mixture. This eliminated the need for air pumps from these engines. Regulations in place for 1970 model automobiles and those later had even tighter restrictions, so Porsche fitted capacitive-discharge ignitions (CDI) to increase spark intensity, further improving exhaust emissions. However, because there was no method to "choke" fuel injection during a cold start, Porsche introduced a hand throttle to set idle during warm-up, placing the control between the seats.

Porsche managed to tweak another 10 horsepower from the base 911 engine. The S stepped up from 160 DIN, 180 SAE to 170 DIN, 190 SAE. To ensure that the additional power output did not strain the S engine, Porsche added an additional oil cooler behind the grille below the right headlight.

Deep inside the engine, Porsche made another profound change. But where fuel injection demonstrated its value, this new decision proved ultimately to be a step backward. Where 911 crankcases from 1964 through mid–model year 1968 had been cast aluminum, Porsche now cast them in magnesium. Threads to secure head bolts were tapped into the cases. These magnesium crankcases saved 22 pounds from the rear of the car, but over the years, most owners would rather have had the extra 22 pounds than the grief the magnesium castings caused.

The engine now used three different metals, magnesium for the crankcase, aluminum for the cylinder heads, and steel for the long bolts holding the heads onto the magnesium casings. Unfortunately these three metals expand at different rates when hot,

with aluminum going the farthest. Over time, this expansion literally pulled the steel bolts from the threads of the magnesium cases. Air pollution controls further heated engines to combust unburned fuel. This caused early engine failures requiring rebuilds at 50,000 to 75,000 miles. Shops drill out the threads, tapping in new ones or installing an insert, and replacing the steel bolts with, most often, bolts made of dilavar. This will have been done at least once to any 1968 magnesium-alloy engine that has 50,000 miles or more.

Porsche improved handling by moving the rear wheels 2.25 inches back, extending the wheelbase to 89.3 inches, without lengthening the car body. By moving the rear wheel aft, the circular access port for the torsion bar changed, and in 1969-and-later models a round plug in the rocker panel provided access and gave a telltale sign making it easy to identify the model year change.

Porsche engineers devised and introduced a self-adjusting suspension. This used hydro-pneumatic front struts developed by Boge that adopted the functions of strut, spring, and anti-sway bar. Offered as standard equipment on 911E models, it was optional on the S and could be ordered on the T models as part of a comfort package that intended to soften the ride. The Boge system definitely softened the ride; its struts provided longer travel than the Koni shock absorbers and MacPherson struts, allowing uncharacteristic body roll in hard cornering.

The hydro-pneumatic struts from Boge soon developed seal leaks and owners found it too costly to replace them. By this point, unless the car has lived on blocks in a museum, there will be no hydro-pneumatic suspension models left anywhere.

Among other smaller changes that improved handling, Porsche changed over battery power from a single 12-volt in the left front corner, to two 6-volt batteries, one below each headlight, to evenly divide the weight up front. To make use of this power, Porsche fitted a 770-watt alternator, which it needed, with the addition of optional air conditioning and a heavily revised heating system.

At the end of model year 1969, Porsche discontinued the 912 series, replacing it with 914 and 914/6 models.

1969 Specifications "B" Series

Body Designation:		911T, 911E, 911S
Price:		911T: $5,795 POE New Jersey
		911E: $6,995
		911S: $7,695 add $620 for Targa
Engine Displacement and Type:		911T: Typ 901/03; 1,991 cc
		(121.5 cid) SOHC Weber 46IPT 3C carburetors, w/EGR Exhaust Gas Recirculation
		911E: Typ 901/09 with Bosch mechanical fuel injection and EGR
		911S: Typ 901/10 with Bosch mechanical fuel injection and EGR
Maximum Horsepower @ rpm:		911T: 125 SAE @ 5,800 rpm
		911E: 158 SAE @ 6,500 rpm
		911S: 190 SAE @ 6,800 rpm
Maximum Torque @ rpm:		911T: 131 ft-lb @ 4,200 rpm
		911E: 145 @ 4,500 rpm
		911S: 152 @ 5,500 rpm
Weight:		911T: 2,360 pounds coupe
		911E: 2,410 pounds coupe
		911S: 2,370 pounds coupe
0–60 mph:		911T: 7.8 seconds (*Car and Driver*)
		911E: 7.0 seconds (*Car and Driver*)
		911S: 6.5 seconds (*Car and Driver*) add 1.2 seconds Sportomatic
Maximum Speed:		911T: 125 mph (*Motor Racing &Sportscar*)
		911E: 130 mph (*Road & Track*)
		911S: 136 mph (*Motor*) subtract 11 mph Sportomatic
Brakes:		ATE Dunlop vented four-piston caliper disc brakes
Steering:		ZF rack-and-pinion
Suspension:	Front:	911T and S: MacPherson struts, telescoping shock absorbers, lower wishbones, longitudinal torsion bars, antisway bar
		911E: wishbones, antisway bar, Boge hydro-pneumatic self-leveling struts
	Rear:	telescoping shock absorbers, semi-trailing arms, transverse torsion bars, antisway bar
Tires:	Front:	911T: 185HR14
		911E, S: 185/70VR15
	Rear:	911T: 185HR14
		911E, S: 185/70VR15
Tire air pressure:	Front:	31 psi; Rear: 34 psi
Transmission(s):		911T: 901/12 4-speed U.S.
		905/13 Sportomatic U.S.
		911E: 901/06 5-speed U.S.
		911S: 901/13 5-speed U.S.
Wheels:	Front:	911T: 5.5Jx14 Fuchs
		911E, S: 6.0Jx15 Fuchs
	Rear:	911T: 5.5Jx14 Fuchs
		911E, S: 7.0Jx15

What they said at the time–Porsche for 1969

Car and Driver, March 1969

"Mark Donohue's first comment on the 911S pretty much said it all. 'Although it's impossible to identify these cars in profile, as soon as you light the fires, there's no doubt which one is the top of the line. That rasping, crackling exhaust note on the 911S really tells you right off it's not fooling around.'

"Porsche would like the world to believe that the primary reason it went to fuel injection was to comply with the exhaust emission standards, but no one is hiding the fact that there's more power available as well. The FI's heart is a six-plunger distributor pump that supplies fuel to each intake port via separate pipes. The flow is intermittent and timed in the firing order sequence."

Parts List for 1969 911s

These are items most commonly replaced during regular maintenance and routine daily operation. Prices quoted are for new factory parts at list price, not including installation labor. NLA means factory parts no longer available, so prices quoted are from aftermarket suppliers.

Engine:

1. Oil filter $16.50
2. Alternator belt $16.00
3. Starter $433.15
4. Alternator (NLA) $899.00
5. Muffler $1,126.98
6. Clutch disc $531.42

Body:

7. Front bumper $1,991.98
8. Left front fender $2,053.12
9. Right rear quarter panel $1,959.18
10. Front deck lid $1,852.54
11. Front deck lid struts .. $43.22 each
12. Rear deck lid struts ... $25.25 each
13. Porsche badge, front deck lid $201.18
14. Taillight housing and lens $854.45
15. Windshield (NLA) $509.00
16. Windshield weather stripping $155.44

Interior:

17. Dashboard (NLA) $1,298.36
18. Shift knob (5-speed) (NLA)..................... $217.47
19. Interior carpet, complete (NLA) $1,800.00

Chassis:

20. Front rotor $205.80
21. Brake pads, front set $96.86
22. Koni rear shock absorber $374.72
23. Front wheel (Fuchs) (NLA)..................... $250–$400
24. Rear wheel (Fuchs) (NLA)..................... $250–$400

Ratings

1969 models, manual transmission

	911T coupe	911E coupe	911S coupe	911T Targa	911E Targa	911S Targa
Acceleration	3	3.5	4	3	3.5	4
Comfort	3	3	4	3	3	4
Handling	3.5	3g	4	3	2.5g	3
Parts	3	3g	4	2b	2bg	2b
Reliability	2f	2fg	2f	2f	2fg	2f

b - Targa roofs no longer available from Porsche.

f - Magnesium cases, see text.

g - Hydro-pneumatic suspension, see text.

Ratings *continued*

1969 models, Sportomatic transmission

	911T coupe	911E coupe	911S coupe	911T Targa	911E Targa	911S Targa
Acceleration	2.5	3	3.5	2.5	3	3.5
Comfort	3	3	4	3	3	4
Handling	3.5	3g	4	2.5	2g	3
Parts	1d	1dg	1d	1bd	1bdg	1bd
Reliability	1ef	1efg	1ef	1ef	1efg	1ef

b - Targa roofs no longer available from Porsche.

d - Sportomatic parts, gaskets extremely hard to find, costly to buy.

e - Sportomatic repair difficult, expensive, may fail quickly. See text.

f - Magnesium cases, see text.

g - Hydro-pneumatic suspension; see text.

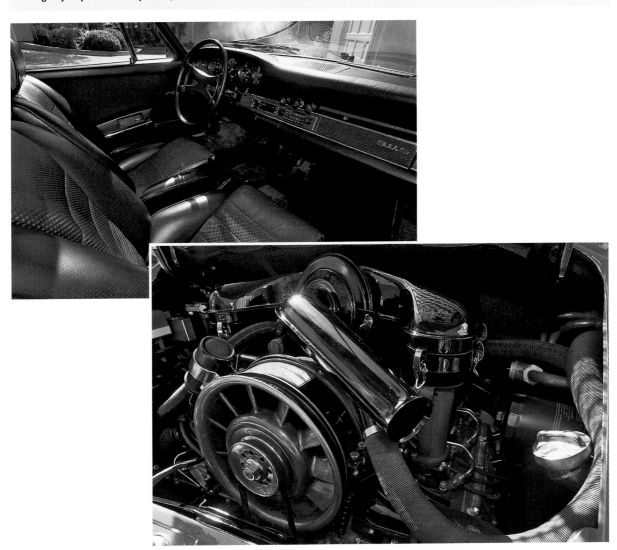

1969 Garage Watch
Problems with (and improvements to) Porsche 911 models.

New longer wheelbase to improve handling, new front suspension, mechanical fuel injection on E and S models, Weber carburetors on T models. Last year of the pull-push type clutch linkage. Axles and CV joints much larger, providing longer life expectancy. Seats and recliner mechanism improved. Hand throttle introduced to aid in cold starting because car has no choke.

Electrically heated rear window defogger.

Timing chain guides/tensioners/tensioner support shaft/idler sprockets all fail, as from 1968.

Cylinder head studs in mag-alloy cases, as from 1968.

Modern day gasoline won't burn fully, as from 1965.

Mechanical fuel injection on E and S models. Tuning difficult. Repair and replacement very costly. Thirsty. System runs well on E and S models and it can be very reliable when set up properly, but in order to run well, they run rich, making it difficult for them now to pass certain state smog tests and providing barely 12-to-15 mile-per-gallon fuel economy.

Typ 905/13 Sporto transmission—parts unobtainable, as from 1968. On cars with 901/12 transmissions as fitted to 911T models, 901/06 transmissions as fitted to 911E models, and 901/13 transmissions in 911S models, the spider gear climbs up into the rest of the transmission, destroying the transmission. As from late 1968.

Last year of push-button outer door handles.

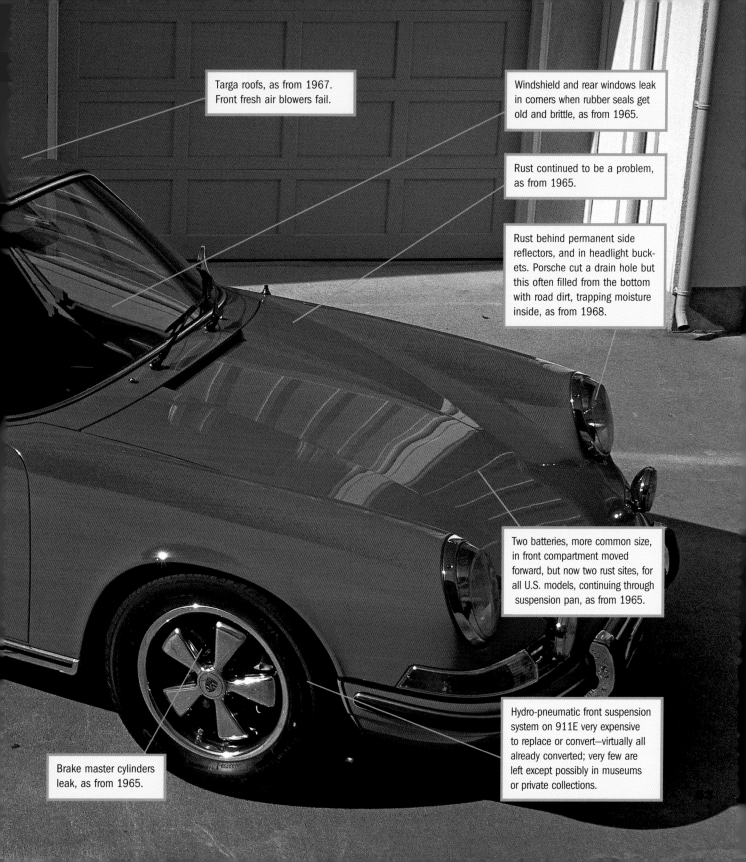

Targa roofs, as from 1967. Front fresh air blowers fail.

Windshield and rear windows leak in corners when rubber seals get old and brittle, as from 1965.

Rust continued to be a problem, as from 1965.

Rust behind permanent side reflectors, and in headlight buckets. Porsche cut a drain hole but this often filled from the bottom with road dirt, trapping moisture inside, as from 1968.

Two batteries, more common size, in front compartment moved forward, but now two rust sites, for all U.S. models, continuing through suspension pan, as from 1965.

Brake master cylinders leak, as from 1965.

Hydro-pneumatic front suspension system on 911E very expensive to replace or convert—virtually all already converted; very few are left except possibly in museums or private collections.

For 1970, Porsche introduced a new 2.2-liter engine, cast in magnesium, as were the final 1,991cc models. Engineers designed a single aluminum cylinder head that worked for the 911S, 911E, and now for 911T models that would be offered worldwide. Horsepower was varied no longer by valve diameter but by timing and, of course, by fuel feed. The T in the United States used Zenith 40TIN carburetors (in Europe it used Webers), while the E and S models continued with the Bosch mechanical injection. The factory affixed a small engine silhouette decal imprinted with "2.2" to the bottom center of each rear window.

Porsche revised the clutch, using what is described as a push-pull type linkage. Depressing the pedal pulled the linkage. Over time, the plastic thrust piece at the rear of the cable will break, leaving you without a clutch. A common fix was to use an aluminum piece, but that pointed up a weak link at the pivot bolt **inside** the bell housing that requires engine removal from the car to replace when it fails. Owners have discovered the best remedy is to use the plastic piece and always carry a spare in the glove box.

The guide hole for the clutch cable is part of a one-piece casting of the differential housing that, in 1970 and later, was magnesium like the crankcases. The holes enlarge over time from use as the cable slides through them. When the hole fails, the clutch cable rises and tends to rub the axle flange. Friction breaks it. A common fix is to attach an auxiliary loop to the outside of the differential housing.

The clutch pressure plate on 1970 and 1971 models resembles that of 1972 and later models. It is not the same and if one tries to use a later plate (at half the price), they'll learn the mechanism is not strong enough to operate the heavier diaphragm.

By now, most throw-out fork pivot bearings are on their third generation of changes and replacements. At some point even these will fail, requiring removal of the engine to replace them.

Porsche continued to offer its semiautomatic Sportomatic, now as Typ 905/20, though this was no longer available for the 911S.

Engineers rearranged hot air vents below the dash to better distribute the heat. The turn signal lever got a midway setting to blink the lights for lane changes. Cars destined for the United States got signal buzzers to remind drivers of keys left in the ignition. Porsche added a lock to the steering column, released by turning the ignition key, as a first step in theft deterrence. The electric rear window defogger got a second intensity level for ice and snow or mist condensation. The factory revised outside door handles again to improve the opener lever and it offered an optional outside rear window wiper, and optional power lift for side windows.

More significant for owners in wet and snowy regions was the introduction of undercoating on all models, as Porsche first addressed rust and corrosion protection. Unfortunately, over the long term, this material had the opposite effect and actually trapped moisture next to metal.

Another limited series bears mention: the rally-prepared cars known as the 911S-T models. These were never really sold to the public and Porsche's racing department prepared a small number, perhaps only nine, of virtually gutted-911T models fitted with modified 911S engines, producing more on the order of 230, 250, and even 270 DIN horsepower. These cars generally weighed around 1,850 pounds, using thin-wall door and deck lid skins similar to those used on the 1967 911R models. They were campaigned vigorously by the factory. Their histories, and their whereabouts, are known and, like the R models, they change hands over the phone for high six-figure prices.

It's fair to say that not all Porsche's improvements always were improvements. But it's important to make clear how good the 1970 model was. Because of the new aluminum cylinder heads with their large valves (at 46 millimeters, they were 1 millimeter larger than the previous 911S heads, and equal to the racing Carrera 6 sizes), any properly running 1970 model ran better, ran faster, and got slightly better mileage than any properly running 1969 model. There was more power, more reliability, more durability, and more civility for the prices.

1970 Specifications "C" Series

Body Designation:		911T, 911E, 911S
Price:		911T: $5,795 POE New Jersey
		911E: $6,995
		911S: $7,695 add $620 for Targa
Engine Displacement and Type:		911T U.S.: Typ 911/07; 2,195 cc (133.9 cid), SOHC, Zenith 40 TIN carburetors w/EGR
		(Exhaust Gas Recirculation)
		911E U.S.: Typ 911/01 with Bosch mechanical fuel injection, EGR
		911S U.S.: Typ 911/02 with Bosch mechanical fuel injection, EGR
911E Maximum Horsepower @ rpm:		911T: 142 SAE @ 5,800 rpm
		911E: 175 SAE @ 6,500 rpm
		911S: 200 SAE @ 6,500 rpm
Maximum Torque @ rpm:		911T: 148 ft-lb @ 4,200 rpm
		911E: 160 @ 4,500 rpm
		911S: 164 @ 5,200 rpm
Weight:		911T: 2,318 pounds coupe
		911E: 2,307 pounds coupe
		911S: 2,390 pounds coupe
0–60 mph:		911T: 8.1 seconds (*Road & Track*)
		911E: 7.7 seconds (*Autosport*)
		911S: 7.3 seconds (*Road & Track*) add 1.2 seconds Sportomatic
Maximum Speed:		911T: 128 mph (*Modern Motor*)
		911E: 137 mph (*Autosport*)
		911E: 144 mph (*Road & Track*) subtract 11 mph Sportomatic
Brakes:		vented disc brakes
Steering:		ZF rack-and-pinion
Suspension:	Front:	911T, S: MacPherson struts, telescoping shock absorbers, lower wishbones, longitudinal
		torsion bars, antisway bar
		911E: hydro-pneumatic self-leveling struts, wishbones
	Rear:	telescoping shock absorbers, semi-trailing arms, transverse torsion bars; S adds antisway bar
Tires:	Front:	911T: 185HR14
		911E, S: 185/70VR15
	Rear:	911T: 185HR14
		911E, S: 185/70VR15
Tire air pressure:	Front:	31 psi; Rear: 34 psi
Transmission(s):		911T: 911/00 4-speed U.S
		911T, E: 905/20 Sportomatic U.S.
		911T, E, S: 911/01 5-speed U.S. Wheels
	Front:	911T: 5.5Jx14 Fuchs
		optional: 5.5J15 Mahle cast aluminum
		911E, S: 6.0Jx15 Fuchs
	Rear:	911T: 5.5Jx14 Fuchs
		optional: 5./5Jx15 Mahle cast aluminum
		911E, S: 7.0Jx15 Fuchs

What they said at the time–Porsche for 1970

Road & Track, March 1970

"The 911S, exciting though it is, is not the right car for use in America—unless one lives in a state without open road speed limits. The S just frustrates its poor driver most of the time in everyday driving, crying to be run up to its redline through the gears (do that in second gear and you're at the speed limit in California) or cruise at 130-plus miles per hour, and gives away a lot in low-speed performance to get its brilliant upper range, not to mention the extra cost. Since the 911T is also a bit stronger this year, and just as tractable as before, we recommend it as the best Porsche for all-around use—and a great car it is."

Parts List for 1970 911s

These are items most commonly replaced during regular maintenance and routine daily operation. Prices quoted are for new factory parts at list price, not including installation labor. NLA means factory parts no longer available, so prices quoted are from aftermarket suppliers.

Engine:

1. Oil filter $16.50
2. Alternator belt $16.00
3. Starter $433.10
4. Alternator (NLA) $899.00
5. Muffler $1,126.98
6. Clutch disc $440.19

Body:

7. Front bumper $1,991.98
8. Left front fender $2,053.12
9. Right rear quarter panel $1,959.18
10. Front deck lid $1,852.54
11. Front deck lid struts .. $43.22 each
12. Rear deck lid struts ... $25.25 each
13. Porsche badge, front deck lid $201.18
14. Taillight housing and lens $848.43
15. Windshield (NLA) $509.00
16. Windshield weather stripping $155.44

Interior:

17. Dashboard $1,298.36
18. Shift knob (5-speed) (NLA) $217.47
19. Interior carpet, complete (NLA) $1,800.00

Chassis:

20. Front rotor $205.80
21. Brake pads, front set $96.86
22. Koni rear shock absorber $374.22
23. Front wheel (Fuchs) (NLA) .... $250–$400
24. Rear wheel (Fuchs) (NLA) .... $250–$400

Ratings

1970 models, manual transmission

	911T coupe	911E coupe	911S coupe	911T Targa	911E Targa	911S Targa
Acceleration	2.5	3	3.5	2.5	3	3.5
Comfort	3	3	4	3	3	4
Handling	4	3g	4	3.5	2.5g	3.5
Parts	3	3g	4	2b	2bg	2b
Reliability	3f	3fg	3f	3f	3fg	3f

b - Targa roofs no longer available from Porsche.

f - Magnesium cases; see text.

g - Hydro-pneumatic suspension; see text.

Ratings continued

1970 models, Sportomatic transmission

	911T coupe	911E coupe	911T Targa	911E Targa
Acceleration	2	2.5	2	2.5
Comfort	3	3	3	3
Handling	4	3g	3.5	2.5g
Parts	1d	1dg	1bd	1bdg
Reliability	2ef	2efg	2ef	2efg

b - Targa roofs no longer available from Porsche.

d - Sportomatic parts, gaskets extremely hard to find, costly to buy.

e - Sportomatic repair difficult, expensive, may fail quickly. See text.

f - Magnesium cases, see text.

g - Hydro-pneumatic suspension, see text.

1970 Garage Watch
Problems with (and improvements to) Porsche 911 models.

Targa roofs, as from 1967.

Weather seal around optional sunroof will leak. Check drains—they need to be open; you can blow them out with compressed air so rain drains to ground.

Electric window lifts optional; check motors and weather seal.

Clutch changed. New transmission, Typ 911/01 eliminates pin bolt wear problem. Factory optional air conditioning fitted under dash.

Mechanical fuel injection on E and S; T now had Zenith 40TIN carburetors.

Cylinder head studs in mag-alloy cases as from 1968.

Timing chain guides/tensioners/tensioner support shaft/idler sprockets all fail, as from 1968.

Check pull-type clutches on 2.2-liter 1970 911s. Plastic thrust piece at rear of cable breaks. The fix was an aluminum piece, but that pointed up a weak link at the pivot bold inside the bell housing that requires engine removal from the car to replace. Pressure plate in 1970 and 1971 cars looks like that of model years 1972 and on. However, do not use the later plate (which is half as expensive) because it is not strong enough to operate the heavier clutch diaphragm and it will fail very quickly.

On early 1970-production cars with 901/12 transmissions as fitted to 911T models, 901/06 transmissions as fitted to 911E models, and 901/13 transmissions in 911S models, the spider gear climbs up into the rest of the transmission, destroying the transmission. As from late 1968.

Typ 905/20 Sportomatic transmission—parts rare, very expensive-to-no-longer-available. This updates original 905/13 but is little improvement.

Rust, still, non-California cars.

Windshield and rear windows leak in corners when rubber seals get old and brittle as from 1965.

Front fresh air blowers fail, as from 1969.

Guide hole for clutch cable is part of one-piece casting of differential housing. Holes wear from use. When it fails, clutch cable rises up and rubs axle flanges, wearing them out and breaking them.

Hydro pneumatic suspension on E, as from 1969; doubtful any are left.

Brake master cylinders leak, as from 1965.

1971

Chapter 10
The "D" Series—
Status Quo
Is Still Good

The U.S. recession that began in 1970 caught up with Germany in 1971. Production dropped to 11,715 cars as prices, buoyed by a disastrous exchange rate from U.S. dollars to German marks, leapt up.

Porsche carried over its new, more powerful magnesium-cast 2.2-liter engines that it introduced the previous model year. The T in the United States continued with the Zenith 40TIN carburetors, and in Europe it retained the Webers, while the E and S models continued with the Bosch mechanical injection. The output range continued from 125 DIN, 142 SAE at 5,800 rpm for the 911T, to 155 DIN, 175 SAE at 6,200 rpm for the 911E, and 180 DIN, 200 SAE at 6,500 rpm for the 911S. The engine silhouette with the "2.2" was still affixed to the rear window.

The transmissions and magnesium differential casings were retained as well. With the 2.2-liter engine, Porsche had revised the clutch for the 1970 model year, using what is described as a push-pull type linkage. Depressing the pedal pulls the linkage. Over time, the plastic thrust piece at the rear of the cable breaks leaving you without a clutch. A common fix was to use an aluminum piece, but that pointed up a weak link at the pivot bolt **inside** the bell housing that requires engine removal from the car to replace. Owners have discovered the best remedy is to use the plastic piece and always carry a spare in the glove box.

The guide hole for the clutch cable is part of a one-piece casting of the differential housing that, in 1970, 1971, and later, was magnesium like the crankcases. The holes enlarge over time from use as the cable slides through them. When the hole fails, the clutch cable rises and tends to rub the axle flange. Friction breaks it. A common fix is to attach an auxiliary loop to the outside of the differential housing.

The pressure plate on 1970 and 1971 models resembles that of 1972 and later models. It is not the same and if one tries to use a later plate (at half the price), they'll learn the mechanism is not strong enough to operate the heavier diaphragm.

By now, decades after manufacture, most throw-out fork pivot bearings are on their third generation of changes and replacements. At some point even these will fail, requiring removal of the engine to replace them.

Porsche continued to offer its semiautomatic Sportomatic, now in third improved designation, as Typ 905/20, though it was still not available for the 911S.

It would be the last year, however, for the Boge hydropneumatic self-leveling suspension. Porsche was learning that the Boge struts, standard on the E models and optional on T and S versions, were failing sooner than standard shock absorbers and they were much more costly to replace. While all 1971 Es were first manufactured with the system (and a build card might indicate an S or T was fitted with it), it would be all but impossible now to find a car that still had the system. When the seals failed, the car settled nearly onto the ground.

Porsche now offered 911S-T models to special private customers though it only produced five for itself. They ordered a 911S and specified option code M490 that, instead of adding equipment, deleted bumper guards, interior sound deadening, carpeting, head lining, and the glove box door. This code fitted lighter-weight, corduroy-covered sport seats, cardboard door trims with strap pulls to open doors, and aluminum bumpers and deck lid. The racing department produced 100 sets of special lightweight components in addition to the M-490 option code, so customers could further lighten their competition cars.

As with earlier factory special cars, the 1971 S-T models that competed and won at Monte Carlo and ran the African Safari are well documented and accounted for and are extremely expensive. Replicas exist, and a number of legitimate M-490 option 911S cars with additional lightweight parts can be found if that's your interest. Otherwise, as mentioned about the 1970 cars, these 1971 models are better, faster, more reliable, more comfortable, and more secure cars than those that came before.

1971 Specifications "D" Series

Body Designation:		911T, 911E, 911S
Price:		911T: $5,795 POE New Jersey
		911E: $6,995
		911S: $8,775 add $620 for Targa
Engine Displacement and Type:		911T U.S.: Typ 911/07; 2,195 cc (133.9 cid), SOHC, Zenith 40 TIN carburetor w/EGR (Exhaust Gas Recirculation)
		911E U.S.: Typ 911/01 with Bosch mechanical fuel injection, EGR
		911S U.S.: Typ 911/02 with Bosch mechanical fuel injection, EGR
Maximum Horsepower @ rpm:		911T: 142 SAE @ 5,800 rpm
		911E: 175 SAE @ 6,500 rpm
		911S: 200 SAE @ 6,800 rpm
Maximum Torque @ rpm:		911T: 148 ft-lb @ 4,200 rpm
		911E: 160 @ 4,500 rpm
		911S: 164 @ 5,500 rpm
Weight:		911T: 2,318 pounds coupe
		911E: 2,307 pounds coupe
		911S: 2,390 pounds coupe
0–60 mph:		911T: 7.9 seconds (*Road & Track*)
		911E: 7.6 seconds (*Road Test*)
		911S: 7.3 seconds (*Road & Track*) add 1.2 seconds Sportomatic (*Modern Motor*)
Maximum Speed:		911T: 128 mph (*Australian Motor Sports*)
		911E: 137 mph (*Road Test*)
		911S: 144 mph (*Road Test*) subtract 11 mph Sportomatic (*Modern Motor*)
Brakes:		vented disc brakes
Steering:		ZF rack-and-pinion
Suspension:	Front:	911T, S: MacPherson struts, telescoping shock absorbers, lower wishbones, longitudinal torsion bars, antisway bar
		911E: hydro-pneumatic self-leveling struts, wishbones
	Rear:	telescoping shock absorbers, semi-trailing arms, transverse torsion bars; S adds antisway bar
Tires: Front:		911T: 185HR14
		911E, S: 185/70VR15
	Rear:	911T: 185HR14
		911E, S: 185/70VR15
Tire air pressure:	Front:	28 psi; Rear: 34 psi
Transmission(s):		911T: 911/00 4-speed U.S.
		911T, E: 905/20 Sportomatic U.S.
		911T, E, S: 911/01 5 speed U.S.
Wheels:	Front:	911T: 5.5Jx14 Fuchs
		optional: 5.5J15 Mahle cast aluminum
		911E, S: 6.0Jx15 Fuchs
	Rear:	911T: 5.5Jx14 Fuchs
		optional: 5./5Jx15 Mahle cast aluminum
		911E, S: 7.0Jx15 Fuchs

What they said at the time—Porsche for 1971

Road & Track, April 1971

"There is no such thing as a nonsporting Porsche. But now that the 911 comes with a choice of three engines and three transmissions, some 911s are more sporting than others and the factory would admit that a 911T with Sportomatic is the least sporting model in the line.... We are inclined to consider the Sportomatic an excellent answer to a question that hasn't been asked. In its favor, it detracts only slightly from the pleasures of driving.

"The benefits are no more overwhelming. Your left foot can stay planted on the floor, true. But there are few traffic situations, say, crawling through town in the rush hour, where the exertion of the clutch pedal becomes annoying or tiring. And the transmission doesn't shift itself, ever. It is not an automatic and thus doesn't lend itself to effortless driving."

Parts List for 1971 911s

These are items most commonly replaced during regular maintenance and routine daily operation. Prices quoted are for new factory parts at list price, not including installation labor. NLA means factory parts no longer available, so prices quoted are from aftermarket suppliers.

Engine:

1. Oil filter $16.50
2. Alternator belt $16.00
3. Starter....................... $433.15
4. Alternator (NLA) $899.00
5. Muffler $1,126.98
6. Clutch disc $440.19

Body:

7. Front bumper............. $1,027.34
8. Left front fender $1,141.65
9. Right rear quarter panel............. $1,048.40
10. Front deck lid $1,852.54
11. Front deck lid struts.. $43.22 each
12. Rear deck lid struts... $59.00 each
13. Porsche badge, front deck lid............. $149.76
14. Taillight housing and lens $848.43
15. Windshield (NLA)....... $509.00
16. Windshield weather stripping $155.04

Interior:

17. Dashboard................. $1,298.36
18. Shift knob (5-speed) (NLA)...................... $217.47
19. Interior carpet, complete (NLA) $1,800.00

Chassis:

20. Front rotor................ $205.80
21. Brake pads, front set $96.86
22. Koni rear shock absorber.......... $374.22
23. Front wheel (Fuchs) (NLA)........................ $250–$400
24. Rear wheel (Fuchs) (NLA)........................ $250–$400

Ratings

1971 models, manual transmission

	911T coupe	911E coupe	911S coupe	911T Targa	911E Targa	911S Targa
Acceleration	3	3.5	3.5	3	3.5	3.5
Comfort	3.5	3.5	4.5	3.5	3.5	4.5
Handling	4	3g	4	3.5	2.5g	3.5
Parts	3	3g	4	2b	2bg	2b
Reliability	3f	3fg	3f	3f	3fg	3f

b - Targa roofs no longer available from Porsche.

f - Magnesium cases, see text.

g - Hydro-pneumatic suspension, see text.

Ratings <small>continued</small>

1971 models, Sportomatic transmission

	911T coupe	911E coupe	911T Targa	911E Targa
Acceleration	2.5	3	2.5	3
Comfort	3.5	3.5	3.5	3.5
Handling	4	3g	3.5	3g
Parts	1d	1dg	0bd	0bdg
Reliability	2ef	2efg	2ef	2efg

b - Targa roofs no longer available from Porsche.

d - Sportomatic parts, gaskets extremely hard to find, costly to buy.

e - Sportomatic repair difficult, expensive, may fail quickly. See text.

f - Magnesium cases, see text.

g - Hydro-pneumatic suspension, see text.

1971 Garage Watch
Problems with (and improvements to) Porsche 911 models.

The 1970 and 1971 models are virtually identical.

Mechanical injection on E and S.

Timing chain guides/tensioners/tensioner support shaft/idler sprockets all fail, as from 1968.

Cylinder head studs in magnesium-alloy cases as from 1968.

Pressure plate in 1970 and 1971 cars, as from 1970.

Hydro-pneumatic suspension on E, as 1968 through 1970; few available then, all replaced by now.

905/20 Sporto transmission parts are rare, expensive; Sportos increasingly rare. This new incarnation is still troublesome.

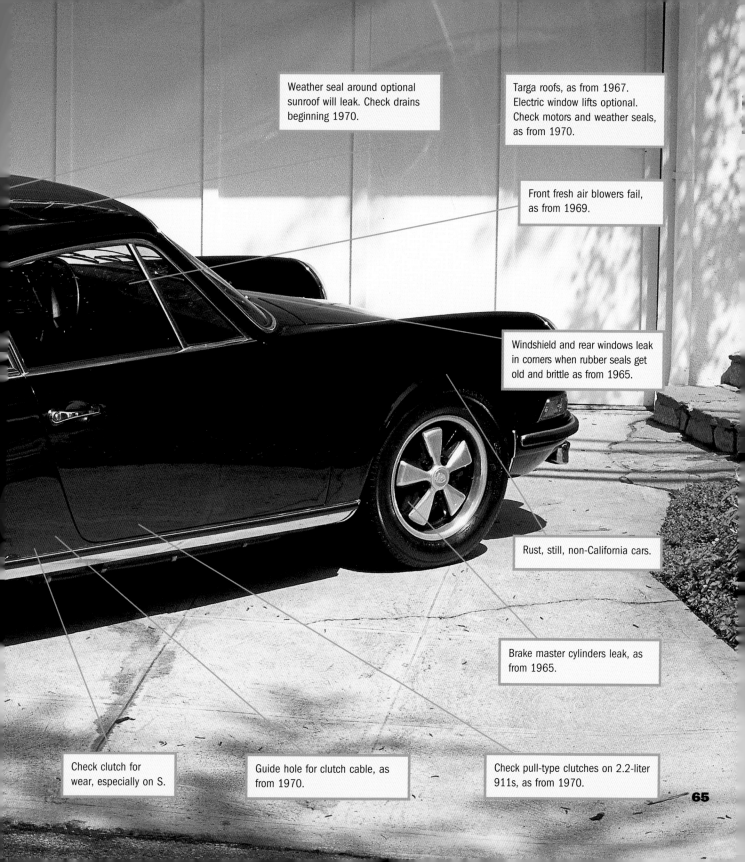

Weather seal around optional sunroof will leak. Check drains beginning 1970.

Targa roofs, as from 1967. Electric window lifts optional. Check motors and weather seals, as from 1970.

Front fresh air blowers fail, as from 1969.

Windshield and rear windows leak in corners when rubber seals get old and brittle as from 1965.

Rust, still, non-California cars.

Brake master cylinders leak, as from 1965.

Check clutch for wear, especially on S.

Guide hole for clutch cable, as from 1970.

Check pull-type clutches on 2.2-liter 911s, as from 1970.

To sell cars in the United States, still home to half the world's Porsche purchasers, Porsche had to clean up engine exhaust emissions further. To do this, they had to decrease the compression ratio in order to run on lower octane "regular" gasoline. This decreased performance and power output as well. However, in order to regain some of what they had sacrificed, they enlarged displacement by increasing stroke to 70.4 millimeters from 66, while retaining the 84 millimeters bore. In order to meet emissions requirements on U.S. models, they dropped the Zenith carburetors from the T and fitted the Bosch mechanical injection. (This led to a kind-of-perplexing designation of 911TV, for rest-of-world cars with Zenith carburetors, and TE models with U.S. "einspritzung." Oddly, the U.S.-only model was more powerful.)

The additional horsepower called for a new transaxle, designated Typ 915. The transmission housing and its rear cover were cast magnesium, while the new differential was die-cast aluminum. Despite using these lightweight raw materials, the new Typ 915 transaxle, albeit much stronger than its predecessor, weighed nearly 20 pounds more.

With this new transmission, Porsche revised its gearshift pattern to a traditional H with first gear above second, unlike earlier five-speed units, which placed first far left below reverse. Porsche argued for years that this was a racing decision as first gear was used only when starting the race. In fact, it had been a design/engineering consideration that required placing first and reverse gears in the cover at the end of the Typ 901/911 and later 916 gearboxes. So many customers complained that Porsche engineers reconfigured the 915 with fifth and reverse tucked in that space, though fifth gear was an extra-cost option and a four-speed gearbox was standard on all models. In addition, to keep performance standards up, engineers lowered overall gear ratios so the more powerful Porsches even seemed quicker. The engineers also strengthened the Sportomatic, now as Typ 925, and a version, 925/01, was offered for the 911S once again.

The 1972 model year saw one oddity that, while it improved weight balance—an argument always championed by engineer Ferdinand Piëch—led to more than a few disasters in service stations. The heavy oil reservoir traditionally sat behind the right rear wheel. Piëch succeeded in placing it ahead of the wheel for model year 1972, and he added an outside filler for it just behind the "B" pillar at the rear of the passenger door. Unfortunately, many unwitting gas station attendants filled the oil reservoir with gasoline, believing the external oil filler was the gas tank and not knowing its filler was under the front deck lid.

While wheel fenders had grown slight flares in previous years as Porsche moved from 4.5-inch wheels to 5.5-inch width, the most striking profile change was the introduction of a lower front air dam or spoiler, standard on 911S models and optional on all others. This wind tunnel–devised fixture reduced front-end lift substantially, making drivers feel much more secure at highway speeds, especially in slippery conditions.

In an attempt to better distinguish the pricey, powerful S from its lesser siblings, especially the E, Porsche revised interior and exterior trim levels. It took the E models down from S level to appear more like the T, though the company still offered these upgrades now as extra-cost options. As a result, there was no 1972 price increase on the E.

In place of the previous hydro-pneumatic suspension on the 911E models, Porsche returned to MacPherson struts and torsion bars as it had before model year 1969 and as T and S cars came standard. In addition, to improve handling and ride, engineers revised the mounting points for rear shocks. Prior to model year 1972, these had tipped slightly to the rear of the rear hubs as they reached the car body. Now they rose straight up and inclined slightly inward.

Porsche now charged $7,367 for the 911T, still $7,995 for the E, and $9,495 for the S at its port-of-entry in New Jersey. The Targa increased to $735, the Sportomatic to $325, and the optional five-speed added $150 to the sticker price.

1972 Specifications "E" Series

Body Designation:		911T, 911E, 911S
Price:		911T: $7,250
		911E: $7,995
		911S: $9,495 add $735 for Targa
Engine Displacement and Type:		911T U.S.: Typ 911/51; 2,341 cc (143.8 cid), SOHC, Bosch mechanical fuel injection w/EGR (Exhaust Gas Recirculation)
		911E U.S.: Typ 911/52 w/Bosch mechanical fuel injection, EGR
		911S U.S.: Typ 911/53 w/Bosch mechanical fuel injection, EGR
Maximum Horsepower @ rpm:		911T: 157 SAE @ 5,600 rpm
		911E: 185 SAE @ 6,200 rpm
		911S: 210 SAE @ 6,500 rpm
Maximum Torque @ rpm:		911T: 166 ft-lb @ 4,000 rpm
		911E: 174 @ 4,500 rpm
		911S: 181 @ 52,00 rpm
Weight:		911T: 2,425 pounds coupe (factory)
		911E: 2,475 pounds Targa (factory)
		911S: 2,455 pounds coupe (factory)
0–60 mph:		911T: 6.9 seconds (*Car and Driver*)
		911E: 6.6 seconds (*Road & Track*)
		911S: 6.0 seconds (*Car and Driver*) add 2.2 seconds Sportomatic (*Modern Motor*)
Maximum Speed:		911T: 129 mph (*Road Test*)
		911E: 139 mph (*Autocar*)
		911E: 145 mph (*Motor*) subtract 9 mph Sportomatic (*Modern Motor*)
Brakes:		vented disc brakes
Steering:		ZF rack-and-pinion
Suspension:	Front:	MacPherson struts with telescoping shock absorbers, lower wishbones, longitudinal torsion bars, anti-sway bar
		911T, E, S: optional hydro-pneumatic self-leveling struts on Boge shocks on E, Koni shocks on S.
	Rear:	telescoping shock absorbers, semi-trailing arms, transverse torsion bars; antisway bar, Boge shocks on E, T, Koni on S
Tires:	Front:	911T: 185HR14
		911E, S: 185/70VR15
	Rear:	911T: 185HR14
		911E, S: 185/70VR15
Tire air pressure:	Front:	28 psi; Rear: 34 psi
Transmission(s):		911T, E: 915/12 4-speed world
		911T, E, S: 915/02 5-speed world
		911T, E: 925/00 Sportomatic U.S.
		911S: 925/01 Sportomatic
Wheels:	Front:	911T: 5.5Jx14 Fuchs
		optional: 5.5Jx15 Mahle cast aluminum
		911E, S: 6.0Jx15 Fuchs
	Rear:	911T: 5.5Jx14 Fuchs
		optional: 5.5Jx15 Mahle cast aluminum
		911E, S: 7.0Jx15 Fuchs

What they said at the time–Porsche for 1972

Road & Track, February 1972

"This is the first Targa we've put through a full test. It's a little bit heavier than the closed coupe and it has more wind noise at speed but otherwise there's little to distinguish it... Any Porsche 911 is very much a driver's car and an experienced and skillful driver can play a 911 as a master violinist plays a Stradivarius. In return for its great potentials of handling, braking, and performance, one makes certain sporting sacrifices of comfort and quiet but no true sports car has ever avoided these and the 911E is as good a compromise as any."

Parts List for 1972 911s

These are items most commonly replaced during regular maintenance and routine daily operation. Prices quoted are for new factory parts at list price, not including installation labor. NLA means factory parts no longer available, so prices quoted are from aftermarket suppliers.

Engine:

1. Oil filter..................... $15.82
2. Alternator belt........... $16.00
3. Starter....................... $433.15
4. Alternator (NLA) $899.00
5. Muffler $1,126.98
6. Clutch disc $440.19

Body:

7. Front bumper............. $1,027.34
8. Left front fender........ $1,141.65
9. Right rear quarter panel............. $1,019.95
(Targa w/outside oil fill) .. $2,356.55
10. Front deck lid............ $1,852.54
11. Front deck lid struts.. $43.22 each
12. Rear deck lid struts... $59.00 each
13. Porsche badge, front deck lid............. $149.76
14. Taillight housing and lens $848.43
15. Windshield (NLA)....... $509.00
16. Windshield weather stripping $155.44

Interior:

17. Dashboard................. $1,298.36
18. Shift knob (5-speed) . $97.42
19. Interior carpet, complete (NLA) $1,800.00

Chassis:

20. Front rotor................. $205.80
21. Brake pads, front set $96.86
22. Koni rear shock absorber.......... $374.22
23. Front wheel (Fuchs) (NLA)...................... $250–$400
24. Rear wheel (Fuchs) (NLA)...................... $250–$400

Ratings

1972 models, manual transmission

	911T coupe	911E coupe	911S coupe	911T Targa	911E Targa	911S Targa
Acceleration	3.5	4	4.5	3.5	4	4.5
Comfort	3.5	3.5	4.5	3.5	3.5	4.5
Handling	4	4	5	3.5	3.5	4.5
Parts	3	3	4	2b	2b	2b
Reliability	3f	3f	3f	3f	3f	3f

b - Targa roofs no longer available from Porsche.
f - Magnesium cases, see text.

1972 models, Sportomatic transmission

	911T coupe	911E coupe	911S coupe	911T Targa	911E Targa	911S Targa
Acceleration	2.5	3	3.5	2.5	3	3.5
Comfort	3.5	3.5	4.5	3.5	3.5	4.5
Handling	4	4	5	3	3	4
Parts	1d	1d	1d	1bd	1bd	1bd
Reliability	2ef	2ef	2ef	2ef	2ef	2ef

b - Targa roofs no longer available from Porsche.   d - Sportomatic parts, gaskets extremely hard to find, costly to buy.
e - Sportomatic repair difficult, expensive, may fail quickly. See text.   f - Magnesium cases, see text.

1972 Garage Watch
Problems with (and improvements to) Porsche 911 models.

Oil tank filler relocated to outside of car body, ahead of right rear wheel. Causes confusion for well-meaning service station attendants and absent-minded owners who filled oil tanks with gasoline.

Outside oil filler. Enlarged engine displacement. First year of pull-pull clutch.

Engines to 2.4 liters.

Cylinder head studs in magnesium-alloy cases from 1968.

915 transmission's second gear synchros, magnesium case. Rear pinion bearing can break out of the cars. First and second gear synchros experienced wear problems. Brake master cylinders leak, as from 1965.

Rust still on non-California cars except around battery box.

Hydro suspension optional on E.

Timing chain guides/tensioners/tensioner support shaft/idler sprockets all fail, as from 1968.

925/00 and 01 Sportomatic transmission, parts rare and expensive. Sportos very rare after 1971. Yet another version, still plagued with troubles.

Targa roofs, as from 1967. Front fresh air blowers fail, as from 1969.

Weather seal around optional sunroof leaks. Check drains from 1970.

Windshield and rear windows leak in corners when rubber seals get old and brittle, from 1965.

Check for clutch condition. If mainshaft seal has failed, clutch contaminates with transmission fluid, requiring rebuild. But bad design in 1972 necessitates removing transmission from car. Modify the case to use 1974 and later guide tube and seal.

In May 1972 Ferry Porsche named Ernst Fuhrmann chairman of the management committee, answering only to Porsche. Fuhrmann had designed the Typ 587 Carrera engine for racing and regular production models but had left Porsche in 1956.

It was not an easy job that Fuhrmann undertook. He had problems to solve and no money to work with. However, his questions and solutions created a legend that redefined the automotive and racing world within a year of his appointment.

The story goes that young racing engineer Wolfgang Berger went with Ernst Fuhrmann to a GT race one weekend shortly after Fuhrmann had returned to Porsche as technical chairman. Porsche 911S models didn't do well. In fact, they were lapped.

"Why are we eating their dust?" Fuhrmann asked his colleague. Berger explained the homologation rules and what Porsche needed to win. Fuhrmann fell silent for some time after taking in Berger's explanation. Then he replied.

"Think it over and let me know what you're going to do about it."

Fuhrmann had been hired to be frugal. When he first visited Weissach, he saw design chief Tony Lapine's drawing boards filled with treatments for front-engined, water-cooled cars, a four-cylinder 924 and a luxurious V-8 engined 928. These too had already consumed money and more was committed. For him, saving money was the objective and to that end, it meant the 911 would have to have many more good years ahead. To make it competitive in racing and keep it desirable in a fickle marketplace, he'd have to breathe new life into Butzi's sleek coupe. The solution

came quickly, a competitive racer with wider rims, the 911S front spoiler and a small rear **Burzel**, or "ducktail" spoiler, a Teldix anti-lock braking system, and a new, highly tuned 2.7-liter engine.

Porsche knew it needed to produce 500 identical cars to qualify for a "production car" category of racing, so company engineers produced a special lightweight edition for homologation, the Carrera RS, for Rennsport or race sport.

The 2.7-liter RS used Mahle's Nikasil process (a nickel silicon carbide coating) for protecting cylinder bores but carried over the Bosch mechanical fuel injection from E and S versions. Porsche won the Group 5 GT championship with the car in September 1972. When Porsche debuted a "production" version at the Paris salon on October 5, it worried how it could sell 500 copies.

For those not wanting to compromise comfort, Porsche offered an M-472 touring package that restored 911S interior trim (and 248 pounds) for $780. Porsche had no plans to offer the car in the United States; its engine was not emissions-legal but without that huge market, would it sell? The lesson of the failed 1967 911R stung, and the sales department rose to the challenge. By the end of the week-long Paris salon, all 500 that Porsche needed for racing qualification had been sold. It immediately offered a second run of 500 (at an additional $312), and eventually added another 800 in both lightweight and touring trim. These nearly 1,800 cars legalized the 911RS for competition not only in Group 4 but in the even less challenging Group 3 where the car would rule.

1973 Specifications "F" Series

Body Designation:		911T, 911E, 911S, 911RS Carrera
Price:		911T: $7,960
		911E: $8,960
		911S: $10,160 add $800 for Targa
		RS Carrera (M471) Sport: $10,200
		RS Carrera (M472) Touring: $11,000
Engine Displacement and Type:		911T U.S.: Typ 911/51; 2,341 cc
		(142.8 cid), SOHC, Bosch mechanical fuel injection w/EGR (Exhaust Gas Recirculation)
		911E U.S.: Typ 911/52 w/Bosch mechanical fuel injection, EGR
		911S U.S.: Typ 911/53 w/Bosch mechanical fuel injection, EGR
		RS Carrera: 911/72, 2,672 cc (162.9 cid), Bosch mechanical fuel injection
Maximum Horsepower @ rpm:		911T: 135 SAE net @ 5 600 rpm
		911E: 159 SAE net @ 6,200 rpm
		911S: 181 SAE net @ 6,500 rpm
		911RS: 200 SAE@ 6,300 rpm
Maximum Torque @ rpm:		911T: 166 ft-lb @ 4,200 rpm
		911E: 174 @ 4,500 rpm
		911S: 181 @ 5,500 rpm
		911RS: 188 @ 5,100 rpm
Weight:		911T: 2,364 pounds coupe (factory)
		911E: 2,425 pounds Targa (factory)
		911S: 2,455 pounds coupe (factory)
		911RS: 1,985 pounds Sport
		911RS: 2,425 pounds Touring
0–60 mph:		911T: 6.9 seconds (*Car and Driver*)
		911E: 6.4 seconds (*Road & Track*)
		911S: 6.0 seconds (*Car and Driver*) add 2.2 seconds Sportomatic (*Modern Motor*)
		911RS: 6.1 seconds
Maximum Speed:		911T: 127 mph (*Motor*)
		911E: 138 mph
		911S: 142 mph (*Road & Track*) subtract 11 mph Sportomatic (*Modern Motor*)
		911RS: 144 mph
Brakes:		vented disc brakes
Steering:		ZF rack-and-pinion
Suspension:	Front:	MacPherson struts with telescoping shock absorbers, lower wishbones,
		longitudinal torsion bars, antisway bar
		911T, E, S: optional hydro-pneumatic self-leveling struts. Boge shocks on E, T., Koni on S
		RS: Bilstein gas-pressurized struts instead of Koni.
	Rear:	telescoping shock absorbers, semi-trailing arms, transverse torsion
		bars; antisway bar, Boge shocks on E, T, Koni on S
		RS: Bilstein gas-pressurized struts instead of Koni
Tires:	Front:	911T: 165/70HR15
		911E, S, 185/70VR15
		RS: 185/70VR15
	Rear:	911T: 165/70HR15
		911E, S, 185/70VR15
		RS: 215/60VR15
	Front:	28 psi; Rear: 34 psi
Transmission(s):		911T: 915/12 4-speed world
		925/20 911T, E, Sportomatic U.S.
		915/02 5-speed
		925/01 911S Sportomatic
		925/08 911RS 5-speed

1973 Specifications "F" Series

Wheels:	Front:	911T: 5.5Jx15 Fuchs steel
		911E: 6.0Jx15 Fuchs steel
		911S, RS: 6.0Jx15 Fuchs alloy
	Rear:	911T: 5.5Jx15 Fuchs steel
		911E: 6.0Jx15 Fuchs steel
		911S, RS: 7.0Jx15 Fuchs alloy

What they said at the time—Porsche for 1973

Road & Track, August 1973
By Joe Rusz

"The year 1973 saw the introduction of a new racing 911—the Carrera. For homologation purposes Porsche built an initial batch of 500 Carrera RS coupes (the road version) and they were snatched up so fast that a second production run is well underway. You can't legally get a Carrera RS for road use in the United States, alas, but it's mouth-watering to consider: lighter than the 911S by over 300 pounds, powered by a 200-horsepower 2.7-liter version of the trusty six . . . and bedecked with a big tail spoiler.

"The Carrera has an even more interesting option list, and part of what's so interesting about it is what doesn't come on the standard model. . . The list goes on to tell us that the RS has less sound insulation than a regular 911 and no bumper moldings, bumper guards, doorsill trim, undercoating, hood springs, glove box or lid, right-hand sun visor, jump seats, clock, dashboard trim, or coat hooks!"

Parts List for 1973 911s

These are items most commonly replaced during regular maintenance and routine daily operation. Prices quoted are for new factory parts at list price, not including installation labor. NLA means factory parts are no longer available, so prices quoted are from aftermarket suppliers.

Engine:

1. Oil filter.................... $15.82
2. Alternator belt........... $16.00
3. Starter...................... $433.15
4. Alternator (NLA)........ $433.15
5. Muffler..................... $1,126.05
6. Clutch disc............... $440.19

Body:

7. Front bumper............. $1,027.34
8. Left front fender........ $1,141.65
9. Right rear
   quarter panel............. $1,019.95

(Targa w/outside oil fill) .. $2,356.55)

10. Front deck lid........... $1,852.54
11. Front deck lid struts.. $43.22 each
12. Rear deck lid struts... $59.00 each
13. Porsche badge,
    front deck lid............ $149.76
14. Taillight housing
    and lens.................... $822.08
15. Windshield (NLA)....... $509.00
16. Windshield weather
    stripping................... $155.44

Interior:

17. Dashboard................ $1,298.36
18. Shift knob (5-speed) . $97.42
19. Interior carpet,
    complete (NLA) $1,800.00

Chassis:

20. Front rotor................ $205.80
21. Brake pads,
    front set.................... $96.86
22. Koni rear
    shock absorber.......... $374.22
23. Front wheel (Fuchs)
    (NLA)........................ $250–$400
24. Rear wheel (Fuchs)
    (NLA)........................ $250–$400

Ratings

1973 models, manual transmission

	911T coupe	911E coupe	911S coupe	911RS-L coupe	911RS-T coupe
Acceleration	3	3.5	4	4.5	4
Comfort	3.5	3.5	4.5	2	4.5
Handling	3.5	3.5	4	5	4
Parts	3	3	4	5	3
Reliability	3f	3f	3f	2f	2f

f - Magnesium engine cases; see text.

1973 models, manual transmission, continued

	911T Targa	911E Targa	911S Targa
Acceleration	3	3.5	4
Comfort	3.5	3.5	4.5
Handling	3	3	3.5
Parts	2b	2b	3b
Reliability	3f	3f	3f

b - Targa roofs no longer available.
f - Magnesium cases, see text.

1973 models, Sportomatic transmission

	911T coupe	911E coupe	911S coupe	911T Targa	911E Targa	911S Targa
Acceleration	2	2.5	3	2	2.5	3
Comfort	3.5	3.5	4.5	3.5	3.5	4.5
Handling	3.5	3.5	4	3	3	3.5
Parts	1d	1d	1d	0bd	0bd	0bd
Reliability	2ef	2ef	2ef	2ef	2ef	2ef

b - Targa roofs no longer available from Porsche.
d - Sportomatic parts, gaskets extremely hard to find, costly to buy.
e - Sportomatic repair difficult, expensive, may fail quickly. See text.
f - Magnesium cases, see text.

1973 Garage Watch
Problems with (and improvements to) Porsche 911 models.

Carrera 2.7 RS models introduced for Europe only. Otherwise, first half-year continued as 1972. Midyear changes substantial.

Electric window lifts optional. Check motors and weather seals, as from 1970.

Front fresh air blowers fail, as from 1969.

First year of all-rubber bumper guards.

Weather seal around optional sunroof leaks, check drain tubes, from 1970.

Windshield and rear windows leak in corners when rubber seals get old and brittle as from 1965.

Rust still, on Midwest, East Coast, or, with battery boxes, in California cars and all others.

Brake master cylinders leak, as from 1965.

U.S. doors get side-beam reinforcement.

Targa roofs, as from 1967.

Timing chain guides/ tensioners/ tensioner support shaft/ idler sprockets all fail, as from 1968.9. 925/00 and 01 Sporto transmissions parts rare and expensive; Sporto rare as option after 1971, as from 1972.

Rotting fuel lines from the tank into the body, down the tunnel at the center of the car then out of the tunnel to the engine cause fuel leaks. There is a fire risk. These leaks can be seen only from underneath the car. If you smell fuel or see a puddle that smells like gas, do not start the car.

Oil tank filler returned to engine compartment location behind rear wheel.

Cylinder head stud inserts, as from 1968.

Magnesium case; second gear synchros remained a concern until the G50 transmission in model year 1987. Midwest and East Coast cars also experience corroded cases if used in winter salt conditions.

First third of production carried over internal mainshaft seal location, as from 1972. Last two-thirds of production year had redesigned transmissions. Rear pinion bearing retainer weak; would break out of the cars. This was fixed by midyear.

Inertia reel seat belts.

Corrosion-resistant muffler improves one rust location.

FEB California

MY 73 RS

1973₁/₂

Chapter 13
The "F" Series
Continued—
The Good Just
Get Better

U.S. emissions and a growing list of safety standards again influenced Porsche design and engineering. While the Piëch-inspired relocation of the oil reservoir to a position ahead of the right rear wheel had led to accidents involving gas stations who filled the tank with gasoline, it had been the risk of accidents on the roads that forced engineers to move the tank back to the rear corner where it had been. U.S. safety regulations began to dictate side impact resistant engineering, and Porsche recognized the exposure of the tank in a broadside crash was too great a risk.

In addition, a legacy of auto insurance lobbying in Washington, D.C., changed the standard of quoting horsepower. By 1972 and 1973, the auto industry recognized that offering products that **seemed** less powerful, even if only on paper, might let them avoid more government regulation. Automakers worldwide began quoting horsepower output in SAE net ratings for U.S. markets, meaning horsepower at the rear tires through the drivetrain, rather than the higher gross figures taken at the flywheel. For Porsche this resulted in down-rating all its engines. Thus, with no engineering changes between 1972 and 1973, the 911S went from 210 SAE gross to 181 SAE net. Likewise, the T dropped from 157 SAE gross to 135 SAE net, and the E slid from 185 gross to 159 net. The new RS 2.7 model provided its fortunate European and Asian buyers with 200 horsepower SAE net.

Porsche added the 1972 911S front spoiler to the standard equipment list on the E model. And another set of U.S. regulations forced large hard-rubber guards onto the bumpers to meet crash recovery standards. Inside the doors, Porsche had to fit side-impact protection beams to reinforce their resistance to incursion into the passenger compartment in collisions.

Meanwhile, catering to American markets, Porsche introduced an electronic fuel injection system fitted to the T. Following its decision to go with mechanical injection for the model year 1969 911E models because of the difficulty of modifying Bosch's electronic version, both Bosch and Porsche engineers continued working with the electronic concept. This new version, named K-Jetronic,

injected fuel in a continuously regulated fashion by using a mainly mechanical metering system supplemented with electrical sensors and overrides. This configuration is referred to as a CIS type, for Continuous Injection System. (The "k" stands for **kontinuierlich**, in German.) Porsche introduced this on its T models.

This system also introduced the airbox and its use came into play only on cold start. The Bosch system automatically pumped a temperature-and-humidity correct mix into the system and this would pass through the box without drama unless the driver mistakenly pumped some gasoline in with the gas pedal. This oversaturated the mix. The first engine spark ignited the mix not only in the cylinders but back to the box. Because this was meant to be a closed system, vacuum pressure was lost and the car would not start or run. Yet drivers felt the need to fuel the engine to aid warm-up, especially on cold mornings. This was the purpose of the hand throttle, which added fuel after the box. With this CIS system, between its introduction in 1973 1/2 and the Bosch introduction of a vacuum operated warm-up regulator in model year 1976, when Porsche did away with the hand throttle, it is imperative never to use the gas pedal on start-up and only to use the hand throttle nestled between the two seats.

To make this CIS system work during normal driving conditions, Paul Hensler had to design new camshafts that made the engine even tamer than previous Ts. Yet with the K-Jetronic managing fuel delivery and spark timing, Hensler held power output at the same gross level as the year before, while increasing torque 3 ft-lb.

The company had increased prices slightly for model year 1973, raising the East Coast port-of-entry charge for a 911T to $7,960. The E had reached $8,960 and the S cracked a big barrier, settling at $10,060. The Targa cost an additional $800. Porsche's most desirable RS 2.7 lightweight commanded $14,000 at the factory door. It was not exported to the United States originally, but many Lightweight and Touring models have subsequently found their way over here.

1973 1/2 Specifications "F" Series

Body Designation:		911T, 911E, 911S, 911RS Carrera
Price:		911T: $7,960
		911E: $8,960
		911S: $10,160 add $800 for Targa
		RS Carrera (M471) Sport: $10,300 (at the factory)
		RS Carrera (M472) Touring: $11,080 (at the factory)
Engine Displacement and Type:		911T U.S.: Typ 911/91; 2,341 cc
		(142.6 cid), SOHC, Bosch K-Jetronic fuel injection
		911E U.S.: Typ 911/52 w/Bosch mechanical fuel injection, EGR
		911S U.S.: Typ 911/53 w/Bosch mechanical fuel injection, EGR
		RS Carrera: 911/72, 2,672 cc (173.7 cid), Bosch mechanical fuel injection
Maximum Horsepower @ rpm:		911T: 134 SAE net @ 5,600 rpm
		911E: 157 SAE net @ 6,200 rpm
		911S: 181 SAE net @ 6,500 rpm
		911RS: 200 SAE net @ 6,300 rpm
Maximum Torque @ rpm (SAE net:		911T: 140 ft-lb @ 4,200 rpm)
		911E: 147 ft-lb @ 4,500 rpm
		911S: 154 ft-lb @ 5,500 rpm
		911RS: 188 ft-lb @ 5,100 rpm
Weight:		911T: 2,364 pounds coupe
		911E: 2,425 pounds coupe
		911S: 2,570 pounds coupe
		911RS: 2,150 pounds Sport
		911RS: 2,398 pounds Touring
0–62 mph:		911T: 7.6 seconds (*Motor*)
		911E: 6.6 seconds
		911S: 7.8 seconds (*Road & Track*) add 1.2 seconds Sportomatic
		911RS: 5.5 seconds (lightweight, factory)
Maximum Speed:		911T: 127 mph (*Motor*)
		911E: 138 mph
		911S: 142 mph (*Road & Track*) subtract 11 mph Sportomatic
		911RS: 149 mph (lightweight, factory)
Brakes:		vented disc brakes
		RS: Teldix anti-lock brake system (ABS)
Steering:		ZF rack-and-pinion
Suspension:	Front:	MacPherson struts with telescoping shock absorbers, lower wishbones, longitudinal torsion bars, antisway bar
		911T, E, S: optional
		hydro-pneumatic self-leveling struts, Boge shocks on E, T, Koni on S
		RS: Bilstein gas-pressurized struts instead of Koni
	Rear:	telescoping shock absorbers, semi-trailing arms, transverse torsion bars, antisway bar, Boge shocks on E, T, Koni on S
		RS: Bilstein gas-pressurized struts instead of Koni
Tires:	Front:	911T: 185HR14
		911E, S: 185/70VR15
		RS: 185/VR7015
	Rear:	911T: 185HR14
		911E, S: 185/70VR15
		RS: 215/60VR15
Tire air pressure:	Front:	28 psi; Rear: 34 psi.

1973 1/2 Specifications "F" Series

Transmission(s):		911T: 915/12 4-speed world
		925/20 911T, E, Sportomatic U.S.
		915/02 5-speed
		925/01 911S Sportomatic
		925/08 911RS 5-speed
Wheels:	Front:	911T: 5.5Jx15 Fuchs steel
		911E: 6.0Jx15 Fuchs steel
		911S: RS: 6.0Jx15 Fuchs alloy
	Rear:	911T: 5.5Jx15 Fuchs steel
		911E: 6.0Jx15 Fuchs steel
		911S: RS: 7.0Jx15 Fuchs alloy

What they said at the time– Porsche for 1973 1/2

Road & Track, August 1973
By Alan Johnson

"The 911S is a hairy, full-bore performance machine, yielding its margins of performance only in the upper reaches of its own capabilities and demanding certain sacrifices, such as low-speed power, fuel economy, and extra money, in return for these margins. Those margins are academic in most of America."

Parts List for 1973 1/2

These are items most commonly replaced during regular maintenance and routine daily operation. Prices quoted are for new factory parts at list price, not including installation labor. NLA means factory parts are no longer available, so prices quoted are from aftermarket suppliers.

Engine:

1. Oil filter..................... $15.82
2. Alternator belt........... $16.00
3. Starter...................... $433.15
4. Alternator (NLA) $899.00
5. Muffler $1,126.98
6. Clutch disc $440.19

Body:

7. Front bumper............. $1,027.34
8. Left front fender........ $1,141.65
9. Right rear
   quarter panel............. $1,019.95
   (Targa w/outside oil fill.... $2,356.55)
10. Front deck lid............ $1,852.54
11. Front deck lid struts.. $43.22 each
12. Rear deck lid struts.... $32.88 each
13. Porsche badge,
    front deck lid............. $149.76
14. Taillight housing
    and lens $822.08
15. Windshield (NLA)....... $509.00
16. Windshield weather
    stripping................... $155.44

Interior:

17. Dashboard................. $1,298.36
18. Shift knob (5-speed) . $97.42
19. Interior carpet,
    complete (NLA) $1,800.00

Chassis:

20. Front rotor................. $205.80
21. Brake pads,
    front set $96.86
22. Koni rear
    shock absorber.......... $374.22
23. Front wheel (Fuchs)
    (NLA .... $250–$400
24. Rear wheel (Fuchs)
    (NLA)......................... $250–$400

Ratings

1973 1/2 models, manual transmission

	911T coupe	911E coupe	911S coupe	911RS-L coupe	911RS-T coupe
Acceleration	3	3.5	4	4.5	4
Comfort	3.5	3.5	4.5	2	4.5
Handling	3.5	3.5	4	5	4
Parts	3	3	4	5	3
Reliability	3.5f	3f	3f	2f	2f

f - Magnesium engine cases; see text.

1973 1/2 models, manual transmission, continued

	911T Targa	911E Targa	911S Targa
Acceleration	3	3.5	4
Comfort	3.5	3.5	4.5
Handling	3	3	3.5
Parts	2b	2b	3b
Reliability	3.5f	3f	3f

b - Targa roofs no longer available from Porsche.
f - Magnesium cases, see text.

1973 1/2 models, Sportomatic transmission

	911T coupe	911E coupe	911S coupe	911T Targa	911E Targa	911S Targa
Acceleration	2	2.5	3	2	2.5	3
Comfort	3.5	3.5	4.5	3.5	3.5	4.5
Handling	3.5	3.5	4	3	3	3.5
Parts	1d	1d	1d	0bd	0bd	0bd
Reliability	2ef	2ef	2ef	2ef	2ef	2ef

b - Targa roofs no longer available from Porsche.
d - Sportomatic parts, gaskets extremely hard to find, costly to buy.
e - Sportomatic repair difficult, expensive, may fail quickly. See text.
f - Magnesium cases, see text.

1973 1/2 Garage Watch
Problems with (and improvements to) Porsche 911 models.

Rust still, as from 1965.

All U.S. cars get Bosch K-Jetronic Continuous (fuel) Injection System. This is a big step forward that nearly doubles gas mileage and significantly improves performance.

Timing chain guides/tensioners/tensioner support shaft/idler sprockets all fail, as from 1968. If guides inside the chain broke, they fell into the chain, derailing it, jamming it, or breaking it, stopping all valve action, as from 1968.

925/00 and 01 Sporto transmissions parts rare and expensive, as from 1968.

Cylinder head studs, as from 1968.

Corrosion resistant muffler improves one rust location.

Airbox on CIS cars. Make a habit of not using gas pedal to start car but instead do use hand throttle (between seats on floor).

First third of production carried over internal mainshaft seal locations, as from 1972. Last two-thirds of production year had redesigned transmissions. Rear pinion bearing retainer weak; would break out of the cars. This was fixed by midyear.

Magnesium case; second gear synchros in 915 transmission. First and second gear synchronizers weak and car must be shifted gently or very expensive transmission failure will be likely. This particular problem was fixed with this model year.

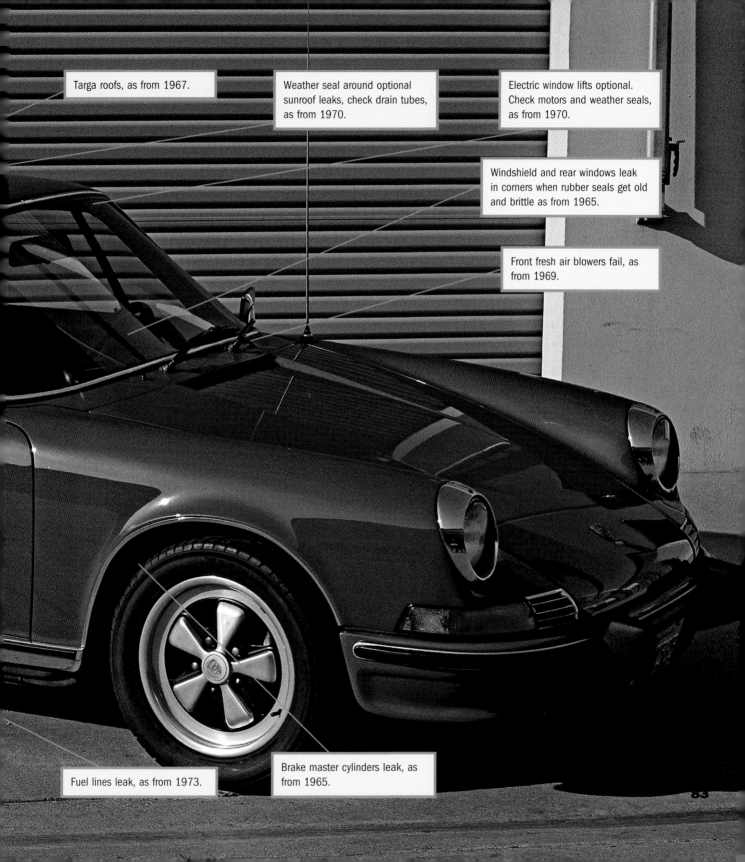

Targa roofs, as from 1967.

Weather seal around optional sunroof leaks, check drain tubes, as from 1970.

Electric window lifts optional. Check motors and weather seals, as from 1970.

Windshield and rear windows leak in corners when rubber seals get old and brittle as from 1965.

Front fresh air blowers fail, as from 1969.

Fuel lines leak, as from 1973.

Brake master cylinders leak, as from 1965.

83

Porsche conceived both the 928 and 924 in answer to U.S. emissions and safety rule-makers who, it seemed to many enthusiasts, were steadily strangling the 911 to death. Yet Porsche proved to skeptics and regulators that engineers love a challenge.

New bumpers appeared on the 1974 models, faired in nicely behind accordion-like rubber housings. New one-piece high-back seats that, when coupled with an optional steering wheel spacer, made tall drivers much more comfortable. The last of the chrome trim disappeared. The transition from chrome to black began in 1973, as outside air intakes, corner light trim, and interior instrument panel bezels went black. Porsche added three-point inertia-reel shoulder and lap belts to the new models.

Designations changed. The T and E models disappeared, replaced by the base 911, the 911S, and a Carrera. But each used the 2.7-liter engine introduced in the 1973 RS. All U.S. cars received the Bosch K-Jetronic (or *kontinuierlich*, continuous injection) engine management system introduced in the midyear 1973 911T. The injection system, along with a thermal reactor, was Porsche's second step in emissions control (the first being a 1968-only air pump). The system relied on extremely high heat to combust any noxious or unburned fuel in the exhaust system. This high heat shortened engine life. A most useful option was M412, the front oil cooler. If the car you're looking at does not have this, be sure to add one as soon as you have acquired the car, especially if you drive in traffic anywhere that air temperatures can regularly exceed 85 degrees Fahrenheit.

Bosch and Porsche continued to develop the CIS system. In its 1973 T application, injection nozzles pumped into the cylinder heads as they had done with the mechanical system in place earlier. However, the electronic system pumped at much lower pressure to keep the mixture lean for better fuel economy and cleaner emissions. But this allowed 1973 models to vapor lock if they were shut off hot. For 1974, Porsche relocated the injection nozzles onto the cast-aluminum inlet manifold, eliminating the vapor lock problem. These improvements and others also slightly increased power output, even for the emissions-strangled U.S. cars. The S and American-market Carrera were reported at 168 SAE net and 174 DIN at 5,800 rpm.

As with the 1973 1/2 models, the 1974 model year 2.7-liter engines used the airbox as the middle air plenum in its air-intake system, so it remained essential to use the hand throttle—and never to touch the gas pedal—on cold start to avoid blowing up the air box. The Bosch system automatically pumped a temperature-and-humidity correct mix into the system and this would pass through the box without drama unless the driver mistakenly pumped some gasoline in with the gas pedal. This would oversaturate the mix. The first engine spark ignited the mix not only in the cylinders but backfiring into the box. Because this was meant to be a closed system, the box exploded, vacuum pressure was lost, and the car would not start or run. Yet drivers felt the need to fuel the engine to aid warm-up, especially on cold mornings. This was the purpose of the hand throttle that added fuel in the runners after the box. With this CIS system, between its introduction in 1973 1/2 and the Bosch introduction of a vacuum operated warm up regulator in model year 1976 (when Porsche did away with the hand throttle), it is imperative never to use the gas pedal on start-up and only to use the hand throttle nestled between the two seats.

Porsche carried over its injection-cast magnesium crank-cases and the Nikasil-plated aluminum cylinders from the 1973 RS for all models. However the much higher temperatures that emissions-strangled engines were generating worried engineers. They developed a new cylinder material, Alusil, an aluminum alloy chemically etched with silicon, and it replaced Nikasil by mid-model year. This extended cylinder life. Porsche also switched to a stainless steel muffler for 1974. But none of this addressed the difference in heat expansion coefficients of magnesium cases and aluminum cylinder heads that pulled the steel cylinder head bolts out of the magnesium castings.

Porsche revised rear suspension pieces for 1974, switching to lighter-weight, less-expensive, and stronger sand-cast aluminum and die-cast magnesium main rear suspension arms.

1974 Specifications "G" Series

Body Designation:		911, 911S, Carrera
Price:		911: $9,950
		911S:,$11,875
		Carrera: $13,575 add $850 for Targa add $425 for Sportomatic
Engine Displacement and Type:		911: Typ 911/92; 2,687 cc (163.9 cid) SOHC, Bosch K-Jetronic fuel injection (CIS), thermal reactor
		911S: Typ 911/93 CIS
		Carrera: Typ 911/72 CIS
Maximum Horsepower @ rpm:		911:143 SAE net @ 5,700 rpm
		911S: 167 SAE net @ 5,800 rpm
		Carrera: 167 SAE net @ 5,800 rpm
Maximum Torque SAE net @ rpm:		911: 168 ft-lb @ 3,800 rpm
		911S: 168 @ 4,000 rpm
		Carrera: 168 @ 4,000 rpm
Weight:		911:2,425 pounds coupe
		911S: 2,443 pounds coupe add 135 pounds for Targa
		Carrera: 2,490 pounds Touring
0–60 mph:		911: 6.1 seconds (*Car and Driver*)
		911S: 5.9 seconds (*Car and Driver*)
		Carrera: 5.8 seconds (*Car and Driver*) add 1.2 seconds Sportomatic
Maximum Speed:		911: 125 mph (*Car and Driver*)
		911S: 129 mph (*Car and Driver*)
		Carrera: 143 mph (*Car and Driver*) subtract 11 mph Sportomatic
Brakes:		ATE/Dunlop vented disk brakes
Steering:		ZF rack-and-pinion
Suspension:	Front:	MacPherson struts with telescoping shock absorbers, lower wishbones,
		longitudinal torsion bars, antisway bar
		Carrera: Bilstein sport shock absorbers
	Rear:	Telescoping shock absorbers, semi-trailing arms, transverse torsionbar; antisway bar
		Carrera: Bilstein shocks
Tires:	Front:	911: 165/HR7015
		911S: 185/70HR15
		Carrera: 185/70VR15
	Rear:	911: 165/HR1570
		911S: 185/70HR15
		911E, S: 215/60VR15
Tire Air Pressure:	Front:	28 psi; Rear: 34 psi
Transmission(s):		911, Carrera: 915/16 4-speed U.S.
		911, S: 925/02 Sportomatic U.S.
		911S: 915/40 5-speed
Wheels:	Front:	911: 5.5Jx15 steel
		911S: 6.0Jx15 ATS "Cookie Cutters"
		Carrera: 6.0Jx15 Fuchs alloy
	Rear:	911: 5.5Jx15 steel
		911S: 6.0Jx15 ATS "Cookie Cutters"
		Carrera: 7.0Jx15 Fuchs alloy

What they said at the time–Porsche for 1974

Autosport, January 1974

By John Bolster

"Superb is a poor word for the suspension, which feels firm and taut, yet soaks up the worst bumps without transmitting a tremor to the occupants. There is none of the pitching that used to be the trademark of rear-engined cars; indeed, the ride is incredibly flat, with very little roll. One typical rear-engined characteristic remains, inasmuch as side winds are annoying at very high speed, though far less so than on the Porsches of a few years ago."

Parts List for 1974 911s

These are items most commonly replaced during regular maintenance and routine daily operation. Prices quoted are for new factory parts at list price, not including installation labor. NLA means factory parts no longer available, so prices quoted are from aftermarket suppliers.

Engine:

1. Oil filter.................... $15.82
2. Alternator belt........... $16.00
3. Starter...................... $433.15
4. Alternator (NLA) $899.00
5. Muffler $890.29
6. Clutch disc $440.19

Body:

7. Front bumper............. $1,949.74
8. Left front fender........ $1,564.86
9. Right rear quarter panel............. $1,730.50
10. Front deck lid $1,865.68
11. Front deck lid struts .. $43.22 each
12. Rear deck lid struts... $59.00 each
13. Porsche badge, front deck lid............. $149.76
14. Taillight housing and lens $822.08
15. Windshield (NLA)....... $509.00
16. Windshield weather stripping................... $155.44

Interior:

17. Dashboard................. $1,298.36
18. Shift knob $97.42
19. Interior carpet, complete (NLA) .... $400–$800

Chassis:

20. Front rotor................. $205.80
21. Brake pads, front set $96.86
22. Koni rear shock absorber $248.00
23. Front wheel (Fuchs) (NLA)....................... $1,158.57
24. Rear wheel (Fuchs) (NLA)....................... $1,158.57

Ratings

1974 models, manual transmission

	911 coupe	911S coupe	911 Carrera coupe
Acceleration	4	4.5	4.5
Comfort	3.5	4	4.5
Handling	3.5	4	4.5
Parts	3	3	3
Reliability	2f	2f	2f

f - Magnesium cases, see text.

1974 models, manual transmission, continued

	911 Targa	911S Targa	911 Carrera Targa
Acceleration	4	4.5	4.5
Comfort	3.5	4	4.5
Handling	3	3.5	4
Parts	2b	2b	2b
Reliability	2f	2f	2f

b - Targa roofs no longer available from Porsche.

f - Magnesium cases, see text.

Ratings continued

1974 models, Sportomatic transmission

	911 coupe	911S coupe	911 Carrera coupe
Acceleration	3.5	4	4
Comfort	3.5	4	4.5
Handling	3.5	4	4.5
Parts	2d	2d	2d
Reliability	2ef	2ef	2ef

d - Sportomatic parts, gaskets extremely hard to find, costly to buy.

e - Sportomatic repair difficult, expensive, may fail quickly. See text.

f - Magnesium cases, see text.

1974 models, Sportomatic transmission, continued

	911 Targa	911S Targa	911 Carrera Targa
Acceleration	3.5	4	4
Comfort	3.5	4	4.5
Handling	3	3.5	4
Parts	2bd	2bd	2bd
Reliability	2ef	2ef	2ef

b - Targa roofs no longer available from Porsche.

d - Sportomatic parts, gaskets extremely hard to find, costly to buy.

e - Sportomatic repair difficult, expensive, may fail quickly, see text.

f - Magnesium cases, see text.

1974 Garage Watch
Problems with (and improvements to) Porsche 911 models.

Seats had headrests for whiplash protection.

Rust, still. Now single battery up front.

Engine to 2.7 in all models, check valve guides.

Airbox: Must do coldstart with hand throttle, as from 1973 1/2.

CIS fuel injection on all models. Improved over original but still touchy, especially on cold starts.

First and second gear synchros as from 1970.

5 mph bumper guards.

Fuel lines leak, as from 1973.

925/02 Sporto trans parts rare, costly.

Brake master cylinders leak, as from 1965.

Standard 911 model has black fan shroud; Carreras (and 911S models) have green fan shrouds. Other problems carried over.

Whale tail appears in Europe; some added after purchase in United States.

Carrera introduced flared rear fenders; uses same engine as 911S. Rear suspension trailing arms now aluminum.

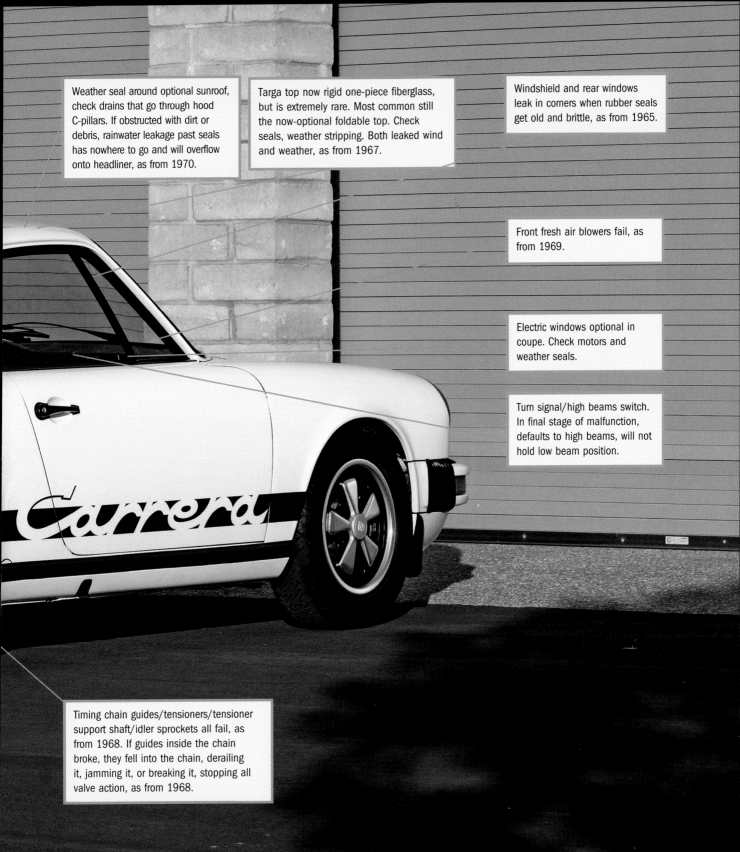

Weather seal around optional sunroof, check drains that go through hood C-pillars. If obstructed with dirt or debris, rainwater leakage past seals has nowhere to go and will overflow onto headliner, as from 1970.

Targa top now rigid one-piece fiberglass, but is extremely rare. Most common still the now-optional foldable top. Check seals, weather stripping. Both leaked wind and weather, as from 1967.

Windshield and rear windows leak in corners when rubber seals get old and brittle, as from 1965.

Front fresh air blowers fail, as from 1969.

Electric windows optional in coupe. Check motors and weather seals.

Turn signal/high beams switch. In final stage of malfunction, defaults to high beams, will not hold low beam position.

Timing chain guides/tensioners/tensioner support shaft/idler sprockets all fail, as from 1968. If guides inside the chain broke, they fell into the chain, derailing it, jamming it, or breaking it, stopping all valve action, as from 1968.

1975

Chapter 15
The "H" Series—
The U.S. EPA Strangles
and California Chokes
the Cars

The world economy continued to sour. Speed limits remained in place and the future for enthusiast drivers looked bleak. Into this mix, the United States Environmental Protection Agency (EPA) and California Air Resources Board (CARB) added further emissions restrictions and the U.S. Department of Transportation (USDOT) began its efforts at protecting careless drivers from themselves.

Porsche, having committed a reluctant Management Chairman Ernst Fuhrmann to carry on the car, met the challenge. Porsche dropped the base 911 but it retained the 911S and the Carrera (and even introduced in Europe a potent 930S turbocharged model). To reinforce its dedication, Ferry Porsche consecrated a Silver Anniversary model. Available in both coupe and Targa configurations, he ordered it painted Diamond Silver metallic only, with a black and silver tweed interior. Porsche shipped half of its total production of 1,500 to the United States.

To meet standards for the 49 states not including California, the engine development crew modified the exhaust flow by using the same manifold and heat exchanger on each side. A belt-driven air pump forced air to the exhaust ports to aid in oxidizing the emissions.

California's tougher limits required two thermal reactors that pumped exhaust to heat exchangers, plumbed to the transverse muffler and from there, forward to the exhaust gas recirculation system (EGR) to reduce nitrogen oxide emissions.

The theory of all these thermal reactors, EGRs, and air pumps was to thoroughly burn not only the raw fuel-air mix but also any unburned noxious substances that made it to the exhaust. This required extremely high temperatures that were very hard on the oil in an air-cooled engine, or more accurately, an oil-cooled engine. As a result, a most valuable option to look for is M412, the front oil cooler. Not having one shortened engine life; if you're looking at a car that does not have one, be sure you add one as soon as you've acquired the car.

As with the 1973 1/2 and 1974 model year cars, the 1975 models used the 2.7-liter engines with the airbox, so it remained essential to use the hand throttle—and never use the gas pedal—on cold start to avoid blowing up the airbox.

Cylinder head studs had different thermal expansion rates. In high operating temperatures, they could literally pull out of the magnesium-alloy cases, requiring rebuild at 80,000 to 120,000 miles in pre-smog cars. Case stud inserts can generally be used at first rebuild; problems only get serious at second rebuild at 200,000 miles. With smog equipment causing higher internal heat, rebuilds started becoming necessary at 50,000 to 75,000 miles. A second one sometimes is not possible at 100,000 to 150,000 miles. This problem began in 1968, now is worse with 1975 emissions control equipment additions. However, this is not a problem if the rebuild is done properly, using dilavar head studs and high-quality stud inserts.

Enthusiasts welcomed one engine-compartment addition. Porsche installed an electric blower to force air through the heating system. Engineers also added sound insulation to the cam covers and, by reducing final drive rations to meet U.S. EPA fuel economy requirements, quieted the engine as well.

Inside the warmer passenger compartment, engineers added more sound insulation and returned the swing-out rear windows they had deleted beginning in model year 1968. To require less work from an air-conditioned engine (thereby reducing emissions), they tinted all the window glass as standard equipment on U.S. cars. They also added a windshield radio antenna and an intermittent rain-mist setting for the wipers. Porsche's previously integrated 2.5 mile-per-hour bumpers now had to meet USDOT 5 mile-per-hour impact standards, requiring the addition of large hard-rubber bumper guards. While these were more obvious to viewers, Styling countered this by painting headlight rims body color on Carreras. This treatment was optional on the 911S, the only other model Porsche imported, having dropped the base 911 for 1975. In further integrating function to form, they anodized the Targa bar flat black, though customers still could order brushed stainless. Lastly, the U.S. Carrera's rear wing grew from the little "burzel," or ducktail, to a full-grown race car "whale tail" wing. (It had been offered as an option on Euro-delivered cars during the previous model year, and many U.S. customers ordered and retrofitted them to 1974 models.)

1975 Specifications "H" Series

Body Designation:		911S, Carrera
Price:		911S: $11,875
		Carrera: $13,475 add $850 for Targa
Engine Displacement and Type:		911S, Carrera (49-states)
		Typ 911/43; 2,687 cc (161.9 cid) SOHC, Bosch K-Jetronic fuel injection (CIS), air pump
		911S, Carrera (California)
		Typ 911/44 CIS, air pump, thermal reactors, exhaust gas recirculation (EGR)
Maximum Horsepower @ rpm:		911S, Carrera:
		(49-states): 157 SAE @ 5,800 rpm
		911S, Carrera:
		(California): 152 SAE @ 5,800 rpm
Maximum Torque @ rpm:		49-state: 166 ft-lb @ 4,000 rpm
		California: 166 @ 4,000 rpm
Weight:		911S: 2,525 pounds
		Carrera: 2,576 pounds
0–60 mph:		911S 49-state: 6.2 (*Car and Driver*)
		Carrera 49-state: 6.2 (*Car and Driver*) add 1.2 seconds Sportomatic
		911S California: 8.2 (*Road & Track*)
		Carrera California: 8.2 (*Road & Track*)
Maximum Speed:		49-state: 132
		California: 134 subtract 11 mph Sportomatic
Brakes:		vented disc brakes
Steering:		ZF rack-and-pinion
Suspension:	Front:	MacPherson struts with telescoping shock absorbers, lower wishbones, longitudinal torsion bars, antisway bar
		Carrera: Bilstein sport shock absorbers
	Rear:	Telescoping shock absorbers, semi-trailing arms, transverse torsion bars; antisway bar
		Carrera: Bilstein shocks
Tires:	Front:	911S: 185/70HR15
		Carrera: 185/70VR15
	Rear:	911S: 185/70HR15
		Carrera: 215/60VR15
Tire air pressure:		Front: 28 psi; Rear: 34 psi.
Transmission(s):		911S, Carrera: 915/45 4-speed U.S. 911S, Carrera: 915/40 5-speed
		911S: 925/40 Sportomatic, world
Wheels:	Front:	911S: 5.5Jx15 ATS Cookie Cutter cast aluminum
		Carrera: 7.0Jx15 Fuchs alloy
	Rear:	911S: 5.5Jx15 ATS Cookie Cutter cast aluminum
		Carrera: 7.0Jx15 Fuchs alloy

What they said at the time–Porsche for 1975

Road & Track, December 1974
By Joe Rusz

"The big change insofar as the new 911s are concerned is the engine. As before it's the same for the S and the Carrera but though displacement remains unchanged at 2,687 cc (Porsche rounds it out at 2.7 liters), power output does not. This year the single engine is rated at 157 SAE net horsepower at 5,800 rpm, 10 horsepower less than last year. This is for the U.S. version, which uses only an air pump for emission control in addition to fuel injection and the usual tailoring of spark timing. The California 911s, which must meet that state's more stringent standards, use not only an air pump but also a pair of thermal reactors in which more mixing and burning can occur. The resultant power loss is more—15 horsepower [from the European version], which results in a SAE net output of 152.

What they said at the time continued

Car and Driver, March 1975

"Skid pad cornering ability (at 0.8G) shows no change from last year's Carrera despite the extra inch of wheel width in the new model, but there does seem to be a minor improvement in road course manners. The tail has learned its place a bit better and it's less inclined to sneak out when you lift off the power at the approach of a turn. Still the difference is small. The new Porsches remain Porsches and must be driven accordingly. More power means more understeer; watch out for lift-throttle oversteer, and be careful when you try to brake hard and turn hard at the same time."

Parts List for 1975 911s

These are items most commonly replaced during regular maintenance and routine daily operation. Prices quoted are for new factory parts at list price, not including installation labor. NLA means factory parts no longer available, so prices quoted are from aftermarket suppliers.

Engine:

1. Oil filter..................... $15.82
2. Alternator belt.......... $16.00
3. Starter...................... $433.15
4. Alternator (NLA) $899.00
5. Muffler..................... $1,347.87
6. Clutch disc............... $440.19

Body:

7. Front bumper............. $1,949.74
8. Left front fender........ $1,564.86
9. Right rear quarter panel............. $1,730.50
10. Front deck lid........... $1,865.68
11. Front deck lid struts.. $43.22 each
12. Rear deck lid struts... $59.00 each
13. Porsche badge, front deck lid............. $149.76
14. Taillight housing and lens $822.08
15. Windshield (NLA)....... $509.00
16. Windshield weather stripping $155.04

Interior:

17. Dashboard................. $1,298.36
18. Shift knob $97.42
19. Interior carpet, complete (NLA) .... $400–$800

Chassis:

20. Front rotor................. $205.80
21. Brake pads, front set $96.86
22. Koni rear shock absorber.......... $213.50
23. Front wheel (Fuchs) (NLA)........................ $1,158.59
24. Rear wheel (Fuchs) (NLA)........................ $1,158.59

Ratings

1975 models, manual transmission (1)

	911S coupe (49-state)	911 Carrera coupe (49-state)
Acceleration	4	4.5
Comfort	4	5
Handling	4	4.5
Parts	3	3
Reliability	2f	2f

f - Magnesium engine cases; see text.

1975 models, manual transmission, continued (2)

	911S coupe (Calif.)	911 Carrera coupe (Calif.)
Acceleration	3	3.5
Comfort	4	5
Handling	4	4.5
Parts	2h	2h
Reliability	2fj	2fj

f - Magnesium cases, see text.
h - Smog control equipment may be required for licensing.
j - Smog controls stress engines, see text.

1975 models, manual transmission, continued (3)

	911S Targa (49-state)	911 Carrera Targa (49-state)
Acceleration	4	4.5
Comfort	4	5
Handling	3.5	4
Parts	3b	3b
Reliability	2f	2f

b - Targa roofs no longer available from Porsche.
f - Magnesium cases, see text.

1975 models, manual transmission, continued (4)

	911S Targa (Calif.)	Carrera Targa (Calif)
Acceleration	3	3.5
Comfort	4	5
Handling	3.5	4
Parts	2bh	2bh
Reliability	2fj	2fj

b - Targa roofs no longer available from Porsche.
f - Magnesium cases, see text.
h - Smog control equipment may be needed for licensing.
j - Smog controls stress engine; see text.

Ratings continued

1975 models, Sportomatic transmission (1)

	911S coupe (49-state)	Carrera coupe (49-state)
Acceleration	3.5	4
Comfort	4	5
Handling	4	4.5
Parts	2d	2d
Reliability	2ef	2ef

d - Sportomatic parts, gaskets extremely hard to find, costly to buy.

e - Sportomatic repair difficult, expensive, may fail quickly. See text.

f - Magnesium cases, see text.

1975 models, Sportomatic transmission, continued (2)

	911S coupe (Calif.)	Carrera coupe (Calif.)
Acceleration	2.5	3
Comfort	4	5
Handling	4	4.5
Parts	2dh	2dh
Reliability	1efj	1efj

d - Sportomatic parts, gaskets extremely hard to find, costly to buy.

e - Sportomatic repair difficult, expensive, may fail quickly. See text.

f - Magnesium cases, see text.

h - Smog control equipment may be needed for licensing.

j - Smog controls stress engine, see text.

1975 models, Sportomatic transmission, continued (3)

	911S Targa (49-state)	Carrera Targa (49-state)
Acceleration	3.5	4
Comfort	4	5
Handling	3.5	4
Parts	2bd	2bd
Reliability	2ef	2ef

b - Targa roofs no longer available from Porsche.

d - Sportomatic parts, gaskets extremely hard to find, costly to buy.

e - Sportomatic repair difficult, expensive, may fail quickly. See text.

f - Magnesium engine cases; see text.

1975 models, Sportomatic transmission, continued (4)

	911S Targa (Calif.)	Carrera Targa (Calif)
Acceleration	2.5	3
Comfort	4	5
Handling	3	3.5
Parts	1bdh	bdh
Reliability	1efj	1efj

b - Targa roofs no longer available from Porsche.

d - Sportomatic parts, gaskets extremely hard to find, costly to buy.

e - Sportomatic repair difficult, expensive, may fail quickly. See text.

f - Magnesium cases, see text.

h - Smog control equipment may be needed for licensing.

j - Smog controls stress engine, see text.

1975 Garage Watch
Problems with (and improvements to) Porsche 911 models.

Turn signal/high beams switch, defaults to high. As from 1974.

Last year of hand throttle (to left of emergency brake) for use during warm-up.

Rust behind permanent side reflectors and in headlight buckets.

Brake master cylinders leak. The brakes would not fully release after being applied. This resulted from pedal bushings stiffening up, or being swollen due to rust.

The flexible portions of the fuel lines from the tank into the body—down the tunnel at the center of the car—then out of the tunnel to the engine, are beginning to rot from age and ozone exposure, causing fuel leaks at high pressure, especially in CIS cars, at about 75 psi.

Engine seal leaks due to emissions heat.

Intermediate shaft for timing chains wore quickly, as from 1974.

First and second gear synchros as from 1972.

Ride height increased to work with 5 mph bumpers.

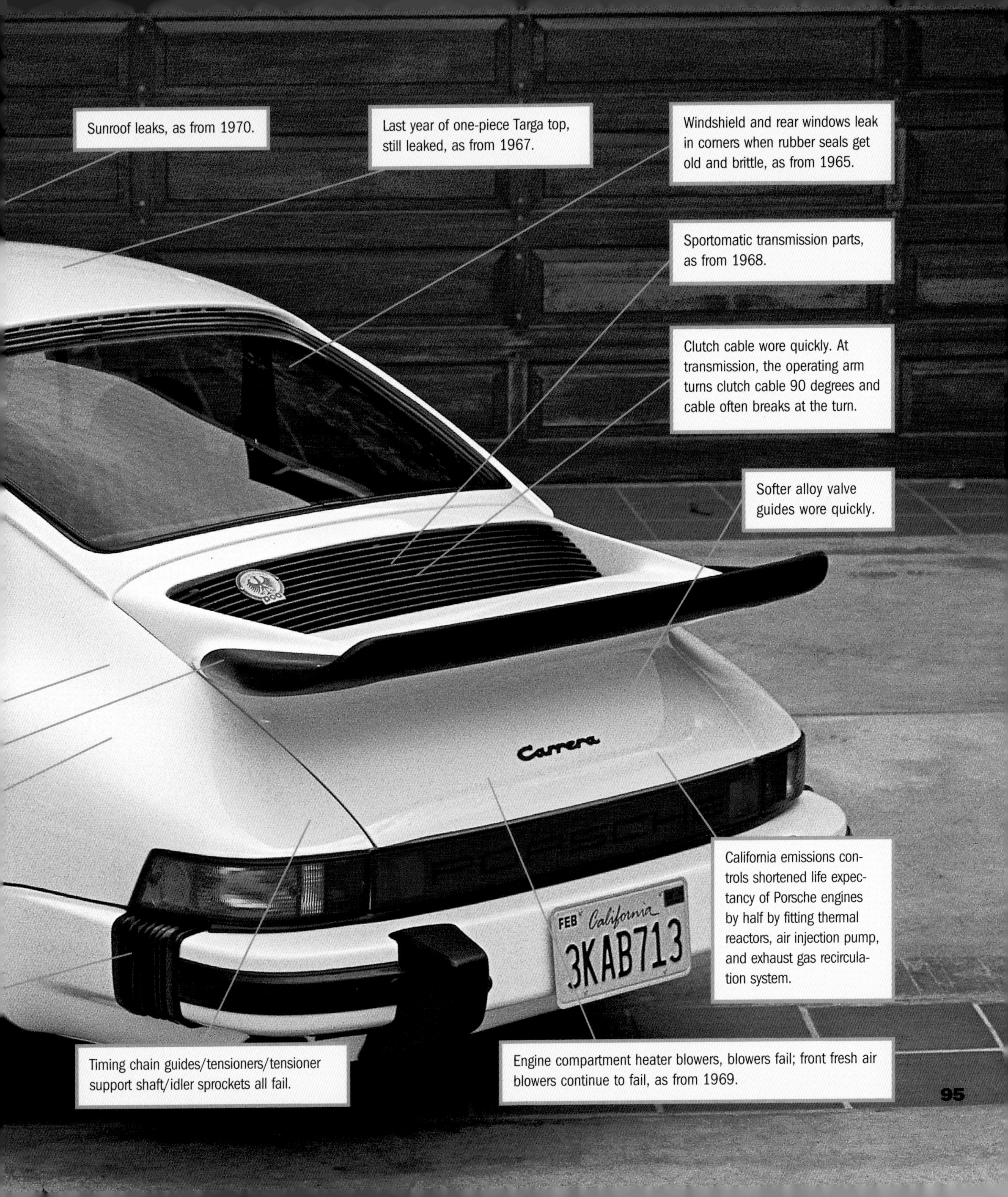

Sunroof leaks, as from 1970.

Last year of one-piece Targa top, still leaked, as from 1967.

Windshield and rear windows leak in corners when rubber seals get old and brittle, as from 1965.

Sportomatic transmission parts, as from 1968.

Clutch cable wore quickly. At transmission, the operating arm turns clutch cable 90 degrees and cable often breaks at the turn.

Softer alloy valve guides wore quickly.

California emissions controls shortened life expectancy of Porsche engines by half by fitting thermal reactors, air injection pump, and exhaust gas recirculation system.

Timing chain guides/tensioners/tensioner support shaft/idler sprockets all fail.

Engine compartment heater blowers, blowers fail; front fresh air blowers continue to fail, as from 1969.

Carrera

FEB California
3KAB713

It was as though nothing else any other carmaker did mattered in any way after Porsche unveiled its turbocharged 930S coupe model at the Paris salon in September 1973 and then announced production in October 1974. Emissions-legal in Europe only, Porsche sold 260 of them starting in March 1975. American customers could only dream, until 1976.

In mid-1975, Porsche began dipping its formed sheet steel body panels through an electrically charged 930-degree bath of zinc. At this temperature, the zinc bonded galvanically to the steel. Every steel surface in the car body was "galvanized" at a manufacturing cost of nearly $100 per car.

Prior to 1976, owners of injected cars did not use the gas pedal but set a hand throttle located between seats to keep engine speed up during warm-up idle. Bosch's K-Jetronic fuel injection system now had a vacuum-operated warm-up regulator with an intake air temperature sensor. This eliminated the need for the hand throttle. Yet not everyone was happy to see it go.

A number of drivers making long autobahn or U.S. interstate drives learned to use the hand throttle as an early cruise control. To substitute that function, Porsche and VDO instruments devised a system—called Tempostat in Europe and Automatic Speed Control for the United States—that held speed anywhere between 25 and 110 miles per hour.

The 1976 models continued with the airbox that Porsche introduced with CIS in 1973 1/2. As before, it was important not to touch the gas pedal during starting or the box could be destroyed, losing vacuum and making the car impossible to start.

One engine compartment change was apparent even to casual observers. Since the 1968 model year, Porsche had used an 11-bladed vertically mounted cooling fan that spun 1.3 times engine speed to force air past the cylinder barrels. In order to run the alternator faster to meet the increased electrical requirements, Porsche changed to a 5-bladed fan that spun at 1.8 times engine speed. This worked great for the alternator but it proved inadequate for engine cooling.

The U.S. buyer had a choice in 1976 of the 911S and the Turbo, both offered in 49-state and California emissions versions. The 911S carried over the 2.7-liter CIS engine.

But it was the Turbo that fueled everyone's fantasies. In 1972, challenged by BMW's 1971 announcement of a production 2002 Turbo, Fuhrmann urged Can-Am engine turbocharging wizard Valentin Schaeffer to prepare a 2.7-liter engine, using the Bosch K-Jetronic injection.

The engine measured 95 millimeters in the bore and used the same 70.4-millimeter stroke as the 2.7-liter engines. Porsche designed airshafts around cylinder head bolts to ensure the long bolts heated and expanded with the cases and heads to avoid the problems that appeared with magnesium cases. Porsche mounted a KKK model 3LDZ turbocharger at the left rear of the engine, easily visible from underneath. It was a tremendously complicated feat of engineering, involving pressure relief valves and interconnected fuel-feed switches. These kept the engine from self-destructing while developing 260 DIN (or 234 SAE net) horsepower at 5,500 rpm, and 253 DIN ft-lb at 4,000. The U.S.-only engine required a thermal reactor and an air pump ahead of the turbo. This spun up the turbo 300 rpm earlier, changing the engine's torque curve. The U.S. version recorded 246 ft-lb SAE net torque at 4,500 rpm. Porsche introduced the 930 S Turbo Carrera to U.S. customers at $25,850, including leather interior, stereo radio, air conditioning, and fog lights standard.

Buyers soon learned the real price—the ability to go far too fast into a corner, lift the gas, and feel the entire car swing like a pendulum. Or, equally often, drivers accustomed to fuel-injected cars would stamp the gas pedal in the middle of a turn. At 2,500 rpm, the turbo boost curve soared vertically to that 234 ft-lb stratosphere, swinging the back end out. And around.

"Finding the trick that reduces throttle lag is the whole goal of our turbocharging work," Fuhrmann told author Jerry Sloniger in mid-1972. For a lot of private owners, that still was a huge problem in 1976.

1976 Specifications "J" Series

Body Designation:		911S, Turbo
Price:		911S: $13,845
		Targa: add $950
		Turbo: $25,850
Engine Displacement and Type:		911S (49-states)
		Typ 911/82; 2,687cc (163.9 cid) SOHC, Bosch K-Jetronic fuel injection (CIS), air pump
		911S (California)
		Typ 911/84 CIS, air pump, thermal reactors, exhaust gas recirculation (EGR)
		Turbo (all U.S.)
		Typ 930/51; 2,994 cc (182.6 cid), BoschK-Jetronic, air pump, thermal reactor, exhaust gas recirculation (EGR)
Maximum Horsepower @ rpm:		911S (49-states):
		157 SAE net @ 5,800 rpm
		911S: (California):
		152 SAE net @ 5,800 rpm
		Turbo: 234 SAE net @ 5,500 rpm
Maximum Torque @ rpm:		911S 49-state: 167 ft-lb @ 4,000 rpm
		911S California: 162 @ 4,000 rpm
		Turbo: 246 @ 4,500 rpm
Weight:		911S: 2,410 pounds
		Turbo: 2,635 pounds (factory)
0–60 mph:		911S: 7.5 (*Road & Track*)
		Turbo: 4.9 seconds (*Car and Driver*)
Maximum Speed:		49-state:add 1.2 seconds Sportomatic
		California: 138 mph subtract 11 mph Sportomatic
		Turbo: 156 mph (*Car and Driver*)
Brakes:		vented disk brakes
Steering:		ZF rack-and-pinion
Suspension:	Front:	MacPherson struts with telescoping shock absorbers, lower wishbones, longitudinal torsion bars, antisway bar
	Rear:	telescoping shock absorbers, semi-trailing arms, transverse torsion bars, antisway bar
Tires:	Front:	911S: 185/70HR15
		Turbo: 185/70VR15
		optional: 205/55VR15
	Rear:	911S: 185/70HR15
		Turbo: 215/60VR15
		optional: 225/50VR15
Tire air pressure:		Front: 28 psi; Rear: 34 psi
	Turbo:	Front: 28 psi; Rear 42 psi
Transmission(s):		911S: 915/44 4-speed U.S.
		911S: 925/04 Sportomatic world
		Turbo: 930/30 4-speed
		Turbo w/50-series tires: 930/32
Wheels:	Front:	911S: 6.0Jx15 ATS Cookie Cutter
		Turbo: 7.0Jx15 Fuchs alloy
	Rear:	911S: 7.0Jx15 ATS Cookie Cutter
		Turbo: 8.0Jx15 Fuchs alloy

What they said at the time–Porsche for 1976

Autosport, July 1975

By Pete Lyons

"With a turbocharged 911 on the autobahn, in daylight hours in the more congested areas, you're on the brakes quite as much as the throttle. Except in bursts, 155 miles per hour is not really a very practical speed.

"I personally found the car impossible to balance, to hold nicely at any steady attitude through a corner. *Rallye Racing* [magazine's] regular test man, far more skilled than I, reached the same conclusion. A couple other people tell me the same. It's a *busy* car."

Parts List for 1976 911s

These are items most commonly replaced during regular maintenance and routine daily operation. Prices quoted are for new factory parts at list price, not including installation labor. NLA means factory parts no longer available, so prices quoted are from aftermarket suppliers.

Engine:

1. Oil filter.................... $18.52
2. Alternator belt........... $16.00
3. Starter...................... $433.15
4. Alternator (NLA) $899.00
5. Muffler (California) ... $1,427.04
   (49-states) $1,347.87
   (Turbo) $2,046.91
6. Clutch disc............... $440.19
   (Turbo) $640.13

Body:

7. Front bumper............. $1,949.74
8. Left front fender........ $1,564.86
9. Right rear
   quarter panel............. $1,730.50
10. Front deck lid........... $1,865.68
11. Front deck lid struts.. $43.22 each
12. Rear deck lid struts... $32.88 each
13. Porsche badge,
    front deck lid............. $149.76
14. Taillight housing
    and lens $822.08
15. Windshield (NLA)....... $509.00
16. Windshield weather
    stripping................... $155.44

Interior:

17. Dashboard................ $1298.36
18. Shift knob $97.42
19. Interior carpet, complete
    (NLA)............... $400–$800
    aftermarket

Chassis:

20. Front rotor................ $205.80
21. Brake pads,
    front set $96.86
22. Koni rear
    shock absorber.......... $213.50
23. Front wheel (Fuchs)
    (NLA)........................ $1,158.59
24. Rear wheel (Fuchs)
    (NLA)........................ $1,158.59

Ratings

1976 models, manual transmission, California and 49-state cars

	911S coupe	Carrera 3.0 coupe	911 Turbo coupe	911S Targa
Acceleration	3	4k	5	3
Comfort	4	4.5k	5	4
Handling	3.5	4k	2f	3
Parts	3	2k	3	2b
Reliability	2f	3k	3	2f

b - Targa roofs no longer available from Porsche.

f - Magnesium cases, see text.

k - Carrera 3.0 1976 model not originally imported but some brought in later.

m - Turbo oversteer, see text.

Ratings <small>continued</small>

1976 models, manual transmission, California and 49-state cars

	911S coupe	Carrera 3.0 coupe	911 Turbo coupe	911S Targa
Acceleration	2.5	4k	4	2.5
Comfort	3	4k	4	3
Handling	3.5	4k	2m	3
Parts	3h	2k	3h	2bh
Reliability	1.5dfj	3k	2.5j	1.5dfj

b - Targa roofs no longer available from Porsche.

f - Magnesium engine cases; see text.

h - Smog control equipment may be needed for licensing.

j - Smog controls stress engine, see text.

k - Carrera 3.0 model not originally imported but some brought in later.

m - Turbo oversteer, see text.

1976 models, Sportomatic transmission (49-state cars)

	911S coupe	Carrera 3.0 coupe	911S Targa
Acceleration	2.5	3.5k	2.5
Comfort	3	4k	3
Handling	3.5	4k	3
Parts	2d	1dk	1d
Reliability	2ef	3ek	2ef

b - Targa roofs no longer available from Porsche.

d - Sportomatic parts, gaskets extremely hard to find, costly to buy.

e - Sportomatic repair difficult, expensive, may fail quickly. See text.

f - Magnesium engine cases; see text.

k - Carrera 3.0 model not originally imported but some brought in later.

1976 models, Sportomatic transmission (California cars)

	911S coupe	Carrera 3.0 coupe (note e)	911S Targa
Acceleration	2	3.5k	2
Comfort	3	4k	3
Handling	3.5	4k	3
Parts	2d	1dk	1bd
Reliability	1.5efj	3ek	1.5efj

b - Targa roofs no longer available from Porsche.

d - Sportomatic parts, gaskets extremely hard to find, costly to buy.

e - Sportomatic repair difficult, expensive, may fail quickly. See text.

f - Magnesium engine cases; see text.

h - Smog control equipment may be needed for licensing.

j - Smog controls stress engine, see text.

k - Carrera 3.0 model not originally imported but some brought in later.

1976 Garage Watch
Problems with (and improvements to) Porsche 911 models.

This is the second of three troubled years that marked the introduction of emissions controls, when engineers made sincere efforts that only exposed other problems. Rust tackled at last with zinc galvanized steel introduced midyear. Rust continues in headlight buckets. Once Porsche removed the glass over the headlight lens, this allowed water to seep behind the beam. Automatic starting system replacing hand throttle. At transmission, the operating arm turns clutch cable 90 degrees and cable often breaks at the turn. This is easy to visually inspect for fraying, as from 1975. Signal/high beam switch, as from 1974.

Introduction of 930 Turbo, wastegate failures. Turbo introduced cast-aluminum crankcase, eliminating stretching and case bolt slippage; aluminum cases also for Carrera to come in 1977 to United States.

Cylinder head stud inserts, as from 1968.

Engine seal leaks due to emission heat, as from 1974.

Five-bladed engine cooling fan, set at a new fan pulley ratio. If owner experiences overheating with this fan, can directly replace this fan with the 1980 11-bladed model.

Timing chain guides/tensioners/tensioner support shaft/idler sprockets all fail.

Airbox, remove engine to replace, as from 1973 1/2.

First and second gear synchros, as from 1972.

Sportomatic transmission parts rare, costly, as from 1968.

Check the limited slip differential on 930 for wear.

The flexible portions of the fuel lines from the tank into the body—down the tunnel at the center of the car—then out of the tunnel to the engine, rot from age and ozone exposure. There is a fire risk. These leaks can be seen only from underneath the car. If you smell fuel or see a puddle that smells like gas, do not start the car, as from 1973.

Targa top two-piece foldable again, still leaks, as from 1967.

Electronic speedometer introduced, allowing Porsche to offer optional M454 Tempostat cruise control. Vacuum leaks possible.

Aftermarket air conditioning compressors; factory air conditioning is through the dash and is poorly ducted. Factory standard on 930 Turbo—also automatic heat control on Turbo. *(NOTE: Air conditioning systems, whether aftermarket or Porsche factory produced, are mediocre at best.)*

California emissions. The extremely high heat required to combust previously unburnt fuel cut engine life to 50,000 to 100,000 miles. This is much less likely if the car was fitted with the factory front oil cooler.

Brake master cylinders leak. This problem disappeared with the introduction of power brakes and its resulting complete redesign of the brake system, but all master cylinders can leak after about 10 years, as from 1965.

Last year of unassisted brakes.

101

1977

Chapter 17
The "J" Series—
The Word for
This Year
Is "Refinement"

As Porsche had done with its RS Carrera, it exceeded even its own hopes with the 1976 Turbo, selling about 1,157 in all, with almost 500 going to the United States. This met the homologation qualifications and allowed for the creation of 31 of the Turbo RSRs as a $40,000 "production" race car. With this, Ernst Fuhrmann ensured an anxious enthusiast world that the 930 would be available again for model year 1977.

As 49-state emissions standards approached California's more strict requirements, Porsche adopted California specifications for all 50 states; yet with improved timing and modifications of the programming for the Bosch K-Jetronic injection, it also recovered the missing five horsepower from 1976 and 1975 models.

The 1977 models continued with the plastic airbox that Porsche introduced with CIS in 1973 1/2. Porsche introduced a backfire relief valve as the factory replacement that allowed excessive pressure to escape without disabling the car. Because the air-boxes are plastic, however, all of them will fail eventually. Still, this CIS system and the airbox represented a big step forward in improving gas mileage and performance.

In model years 1977 through 1979, Porsche began having problems with the zinc galvanizing process and in some areas the zinc (supposed to be between 10 microns and 50 microns thick) ended up much thicker, and it didn't hold paint well.

To lessen the risk of break-in, Porsche devised a roller door lock mechanism inside the door rather than the sill-mounted buttons. For the same security reasons, it also eliminated opening front quarter-windows on both driver's and passenger's doors. This adversely affected air ventilation, so it redesigned the dashboard to allow room for two large fresh air and air-conditioning vents. While this was an improvement, it was poorly ducted. Because Stuttgart is located at a latitude approximately equal to Seattle, Washington, air conditioning was much less necessary than it might be to buyers in Southern California. A Porsche air conditioning authority, Scott Hendry at Performance Aire in Anaheim, California, offers a kit that adds two additional large vents below the dash to nearly double the airflow.

Both U.S. models—the 911S and Turbo—got vacuum power brake boosters. Early owners of 930 S models complained about the very high pedal pressure required to slow or stop their heavier Turbos. This alleviated that complaint. Turbos also got a pressure boost gauge inset within the tachometer.

To improve ride and handling, Porsche engineers stepped up the tire and wheel designations for the Turbo, going from 15-inch wheels to 16s, with 7-inch front and 8-inch rear rim widths to accommodate 205/55VR rated tires in front and 225/50VR tires in back.

Yet because other U.S. customers complained about ride harshness and noise, Porsche also devised a $495 option, M-590, the Comfort Kit. With this, Porsche mounted the car on 14-inch HR-speed rated tires and wheels and electronically governed the car to 130 miles per hour maximum. *Car and Driver* magazine, in its usual pithy cynicism, dubbed it the Porsche Brougham.

Porsche celebrated a milestone in July when it produced its 250,000th automobile. Of the nearly 170,000 911s produced between 1964 and then, slightly more than half had been sold in the United States.

It is worth considering that at this point the Porsche 911 was 14 years old. Engineering had gone far with the car from its original 1,991cc carbureted engine churning out 148 SAE gross horsepower with no regard for what came out of the exhaust pipe. Porsche had squeezed 234 SAE net horsepower, something more like 275 gross, out of just 50 percent more engine size than the 1965 model, while what came from the exhaust pipe satisfied an exceedingly skeptical California Air Resources Board charged with keeping California beautiful. Weissach's engineers did all this while lengthening the wheelbase by 2.4 inches and widening the track from 1,367 millimeters front and 1,335 millimeters rear to 1,369 front and a vast 1,511 millimeters rear for the Turbo. Yet with the exception of a front "chin spoiler" and the Turbo's large, flat whale tale, they did not have to modify the body of the car at all to accommodate this near doubling of power and performance. That engine compartment was getting pretty full, with a turbocharger, air pump, thermal reactors, exhaust gas recirculation pump, and heater fan.

1977 Specifications "K" Series

Body Designation:		911S, Turbo
Price:		911S: $13,845
		Targa: add $950
		M590 Comfort option: $495
		Turbo: $28,000
Engine Displacement and Type:		911S (all 50 states)
		Typ 911/85; 2,687 cc (163.9 cid) SOHC, Bosch K-Jetronic fuel injection
		(CIS), air pump, thermal reactors, EGR
		Turbo (all U.S.)
		Typ 930/53; 2,994 cc (182.6 cid), Bosch
		CIS, air pump, thermal reactors, EGR
Maximum Horsepower @ rpm:		911S: 50-state:
		157 SAE net @ 5 800 rpm
		Turbo: 234 SAE net @ 5,500 rpm
Maximum Torque @ rpm:		911S 50-state: 168 ft-lb @ 4,000 rpm
		Turbo: 246 @ 4,500
Weight:		911S: 2,470 add 90 pounds for Targa
		Turbo: 2,785 pounds
0–60 mph:		911S: 7.5 seconds (factory)
		Turbo: 5.7 seconds (factory) add 1.2 seconds Sportomatic
Maximum Speed:		50-state: 134 mph (factory) subtract 11 mph Sportomatic
		optional: 130 mph w/M590
		Comfort Package
		Turbo: 152 mph (factory)
Brakes:		vented disc brakes
		optional: w/M590 Comfort Package and Sportomatic, power-assisted brakes
Steering:		ZF rack-and-pinion
Suspension:	Front:	MacPherson struts with telescoping shock absorbers, wishbones,
		torsion bars, antisway bar, Optional softer Bilstein shocks in M590 Comfort Package
	Rear:	telescoping shock absorbers, two-piece trailing arms, transverse
		torsion bars, antisway bar, Optional softer Bilstein shocks in M590 Comfort package.
Tires:	Front:	911S: 185/70HR15
		Optional: 185HR14 w/M590 Comfort Package
		Turbo: 205/55VR16
	Rear:	911S: 185/70HR15
		Optional: 185HR14 (with M590)
		Turbo: 225/50VR16
Tire air pressure:	Front:	28 psi; Rear: 34 psi
		Turbo: Front: 28 psi; Rear 42 psi
Transmission(s):		911S: 915/61 4-speed U.S.
		911S: 925/17 Sportomatic U.S.
		Turbo: 930/33 4-speed
Wheels:	Front:	911S: 6.0Jx15 ATS "Cookie Cutter"
		Optional: 6.0Jx14 Fuchs alloy, with M590
		Turbo: 7.0Jx16 Fuchs alloy
	Rear:	911S: 7.0Jx15 ATS "Cookie Cutter"
		Optional: 6.0Jx14 Fuchs alloy with M590
		Turbo: 8.0Jx16 Fuchs alloy

What they said at the time—Porsche for 1977

Road Test, November 1977

"It's been said in the past, of many engines, that they can be fired up. It implies an instant response to the starter and a whomping bellow of sound and usually applies only to racing engines. The air-cooled, fuel-injected flat-six of the 911S fires *right* up and sounds more like its racing derivatives than any other current production engine. Wind it up and it shrieks the good and willing sound of easy power. The 911S rates a genuine 10 on the scale of good sounds."

Parts List for 1977 911s

These are items most commonly replaced during regular maintenance and routine daily operation. Prices quoted are for new factory parts at list price, not including installation labor. NLA means factory parts no longer available, so prices quoted are from aftermarket suppliers.

Engine:

1. Oil filter..................... $18.52
2. Alternator belt........... $16.00
3. Starter....................... $433.15
4. Alternator (NLA) $899.00
5. Muffler (California) ... $1,427.04
   (49-states) $1,347.87
   (Turbo) $2,046.91
6. Clutch disc............... $440.19
   (Turbo) $640.13

Body:

7. Front bumper............. $1,949.74
8. Left front fender........ $1,564.86
9. Right rear
   quarter panel............. $1,730.50
10. Front deck lid $1,865.68
11. Front deck lid struts.. $43.22 each
12. Rear deck lid struts... $32.88 each
13. Porsche badge,
    front deck lid............. $149.76
14. Taillight housing
    and lens $822.08
15. Windshield (NLA)....... $470.06
16. Windshield weather
    stripping................... $155.44

Interior:

17. Dashboard................. $1,298.36
18. Shift knob $97.42
19. Interior carpet, complete
    (NLA)......................... $400–$800
    aftermarket

Chassis:

20. Front rotor................. $205.80
21. Brake pads,
    front set $96.86
22. Koni rear
    shock absorber.......... $213.50
23. Front wheel (Fuchs)
    (NLA)......................... $1,158.59
24. Rear wheel (Fuchs)
    (NLA)......................... $1,158.59

Ratings

1977 models, manual transmission (49-state cars)

	911S coupe	911 Carrera 3.0 coupe	911 Turbo coupe
Acceleration	3	4k	5
Comfort	4	4.5k	4
Handling	3.5	4k	2m
Parts	3	2k	3
Reliability	2f	3k	3

f - Magnesium cases, see text.
k - Carrera 3.0 model not originally imported but some brought in later.
m - Turbo oversteer, see text.

1977 models, manual transmission (California cars)

	911S coupe	911 Carrera 3.0 coupe	911 Turbo coupe
Acceleration	2.5	4k	4
Comfort	4	4.5k	5
Handling	3.5	4k	2m
Parts	3h	2hk	3h
Reliability	1.5fj	3k	2.5j

f - Magnesium engine cases; see text.   h - Smog control equipment may be needed for licensing.   j - Smog controls stress engine, see text.
k - Carrera 3.0 model not originally imported but some brought in later.   m - Turbo oversteer, see text.

Ratings continued

1977 models, manual transmission, continued (49-state cars)

	911S Targa	911 Carrera 3.0 Targa
Acceleration	3	4k
Comfort	4	4.5k
Handling	3	3.5k
Parts	3b	1bk
Reliability	2f	3k

b - Targa roofs no longer available from Porsche.
f - Magnesium engine cases; see text.

1977 models, manual transmission, continued (California cars)

	911S Targa	911 Carrera 3.0 Targa
Acceleration	2.5	4k
Comfort	4	4.5k
Handling	3	3.5k
Parts	3bh	1bhk
Reliability	1.5fj	3k

b - Targa roofs no longer available from Porsche.
f - Magnesium cases, see text.
h - Smog control equipment may be needed for licensing.
j - Smog controls stress engine, see text.

1977 models, Sportomatic transmission (49-state cars)

	911S coupe	911 Carrera 3.0 coupe	911S Targa	911 Carrera 3.0 Targa
Acceleration	2.5n	3.5kn	2.5n	3.5kn
Comfort	4	4.5k	4	4.5k
Handling	3.5	4k	3	3.5k
Parts	2d	2dk	1bd	1bdk
Reliability	1.5ef	2ek	1.5ef	2ek

b - Targa tops no longer available from Porsche.
d - Sportomatic parts, gaskets extremely hard to find, costly to buy.
e - Sportomatic repair difficult, expensive, may fail quickly. See text.
f - Magnesium engine cases; see text.
k - Carrera 3.0 not originally sold in U.S.; some have been brought in.
n - Sportomatic now three-speed, see text.

1977 models, Sportomatic transmission (California cars)

	911S coupe	911 Carrera 3.0 coupe	911S Targa	911 Carrera 3.0 Targa
Acceleration	2n	3kn	2n	3kn
Comfort	4	4.5k	4	4.5k
Handling	3.5	4k	3	3.5k
Parts	2dh	2dhk	1bdh	1bdhk
Reliability	1efj	1.5ejk	1efj	1.5ejk

b - Targa tops no longer available from Porsche.
d - Sportomatic parts, gaskets extremely hard to find, costly to buy.
e - Sportomatic repair difficult, expensive, may fail quickly. See text.
f - Magnesium cases, see text.
h - Smog control equipment may be needed for licensing.
j - Smog controls stress engines, see text.
k - 3.0 Carrera models not originally sold in U.S., some have been brought in.
n - Sportomatic now three-speed, see text.

1977 Garage Watch
Problems with (and improvements to) Porsche 911 models.

This is the third of three years marking the introduction of—and surrender to—emissions control devices. Avoid extremely low mileage original owner cars. They will require much work to remedy existing conditions. If the work already has been done (and any car with more than 50,000 miles will have this work done), verify the quality of the work done to repair, replace, and remedy these problems and parts. Such cars are likely to be a reliable—and very affordable—choice.

Timing chain guides/tensioners/tensioner support shaft/idler sprockets all fail, as from 1968.

Last of 2.7-liter magnesium-alloy blocks, eliminates problem of rebuilds requiring head bolt inserts, as from 1968.

Last of thermal reactors though most removed long ago.

Airbox, remove engine to replace, as from 1973 1/2.

First and second gear synchros, as from 1972.

Brake master cylinders leak. This problem disappeared with the introduction of power brakes and its resulting complete redesign of the brake system for model year 1977, but all master cylinders can leak after about 10 years, as from 1965.

First-year vacuum-boosted brakes.

Sunroof leaks, as from 1970.

Targa roofs leak wind noise and water. Seals are very costly and hard to find, as from 1967.

Windshield and rear windows leak in corners when rubber seals get old and brittle, as from 1965.

Turbo boost gauge on tach.

Heater blowers fail, as from 1975; front fresh air blowers continue to fail, as from 1969.

Sixteen-inch wheels and P7 tires on 930, improving performance and handling greatly.

Incorrect wheels. Fuchs not polished.

Aftermarket air conditioning compressors; factory air conditioning is through the dash and is poorly ducted. Factory standard on 930 Turbo—also automatic heat control on Turbo. (NOTE: Air conditioning systems, whether aftermarket or Porsche factory produced, are mediocre at best.)

Optional Comfort Package provided electric windows and 130 mph speed governor for S. Watch window seals and motors.

The flexible portions of the fuel lines from the tank into the body—down the tunnel at the center of the car—then out of the tunnel to the engine, rot from age and ozone exposure. There is a fire risk. These leaks can be seen only from underneath the car. If you smell fuel or see a puddle that smells like gas, do not start the car, as from 1973.

Typ 925 Sportomatic transmissions parts rare, costly, as from 1968.

It had long been a matter of pride with Porsche's engineers at Weissach to beat the rules makers at their own games in racing, and to beat the rules makers at their own deadlines for regular production models. With the new 3.0-liter SC model, Porsche made a serious effort at anticipating worldwide emissions concerns. Their solution simplified Porsche's product line-up by installing the U.S.-required air pump on European models. There still were differences, however. The U.S. cars for all 50 states got catalytic converters and the California cars still used an EGR system. (For California owners facing emissions tests semiannually, or for buyers needing the test to transfer title, so long as the air pump, EGR, and catalytic converter are operational, this car can pass California Smog tests easily. If the air pump has been disconnected, it clogs up and it cannot simply be rebelted up to pass smog tests. In the late 1970s and early 1980s, California owners believed they attained better performance without the air pump. Initially this was true but over time, performance deteriorated without the system in working order.)

With the worldwide utilization of the 3.0-liter Carrera engine, Porsche had discontinued the magnesium crankcases of the 2.7-liter engines. Reverting to an aluminum crankcase and aluminum heads eliminated the risk of steel case head bolts pulling out of casings due to differing coefficients of expansion from engine heating.

A not so successful revision was a new clutch with a large rubber damper in the center. Porsche intended that this would greatly diminish the vibration and noise of the transmission. Unfortunately, the rubber centers were not capable of handling sustained high fluid temperatures and they disintegrated, usually lodging a piece of the rubber between the clutch and the pressure plate, jamming the clutch. While Porsche continued these rubber center clutches through model year 1980, it is very unlikely any still remain in cars.

The three-speed Sportomatic was becoming less and less popular among U.S. car buyers and for model year 1978, Porsche made the transmission a special order option only (M09) and removed it from its regular catalogs in preparation for phasing it

out entirely. (It would be gone by model year 1980.) A five-speed transmission was standard for SC models, though the Turbo still was available only with four gears.

Porsche introduced the Bosch breakerless capacitive discharge ignition (CDI) on 1978 SC and Turbo models. But the new style pointless distributors rotated in the opposite direction from all previous 911 engines. Therefore what appears to be a spark retard on this distributor (and on the 1979 model year cars) can only advance timing. Still, this new ignition system helped prevent the backfires that blew up the airboxes.

Turbo models gained the biggest and most noticeable change beginning with the 1978 model year. Porsche added an air-to-air intercooler to intensify the fuel-air charge into the engine. Porsche laid this radiator-looking device across the top of the engine fan like a fine-mesh tray below the rear wing. Porsche also increased engine displacement, enlarging bore from 95 to 97 millimeters and stroke from 70.4 to 74.4 millimeters, for a total engine capacity of 3,299 cc, 201.2 cid. This increased horsepower output to 300 DIN, 265 SAE net at 5,500 rpm. Torque rose to 290 ft-lb at 4,000 rpm. In order to meet emissions standards, Porsche fitted a distributor with a double vacuum unit that retarded timing at full throttle. This cost California Turbo models an additional 12 horsepower.

To better stop the Turbo models, Porsche fitted "floating" cross-drilled brake rotors and four-piston calipers derived from the 917 racing models. Porsche had experimented with these cross-drilled discs earlier but had experienced cracking problems from inadequate hardening. With this new system, owners in salt-air environments complained of corrosion problems with the rotors. Porsche converted to nonfloating rotors in the later SC models and many owners have paid for this conversion themselves on 1978 and 1979 Turbo models.

Lastly, to keep the Turbo glued to the ground, Porsche revised the design of the rear wing, from the whale tail to the "tea tray" type of wing with upturned rubber rimmed edges to better manage airflow.

1978 Specifications "L" Series

Body Designation:		911SC, Turbo
Price:		911SC: $17,950
		Targa: $19,050
		Turbo: $36,700
Engine Displacement and Type:		911SC (49-states)
		Typ 930/04; 2,994 cc (182.6 cid) SOHC, Bosch K-Jetronic fuel injection (CIS), air pump, thermal reactors, Catalytic converters
		911SC (California)
		Typ 930/06; 2,994 cc (182.6 cid) SOHC, Bosch K-Jetronic fuel injection (CIS), air pump, thermal reactors, EGR, Catalytic converters
		Turbo (49-state)
		Typ 930/61; 3,299 cc (201.2 cid), Air-to-air intercooler, Bosch CIS, Air pump, thermal reactors, EGR
		Turbo (California)
		Typ 930/63; 3,299 cc (201.2 cid), air-to-air intercooler, Bosch CIS air pump, thermal reactors, EGR, vacuum control ignition retard
Maximum Horsepower @ rpm:		911SC: 49-states:
		180 SAE net @ 5,500 rpm
		California: 172 SAE net @ 5,500 rpm
		Turbo: 265 SAE net @ 5,500
		California: 253 SAE net @ 5,500
Maximum Torque @ rpm:		911SC: 49 states
		187 ft-lb @ 4,200 rpm
		California: 175 @ 4,200
		Turbo: 290 @ 4,000
		California: 282 @ 4,000
Weight:		911SC: 2,558 pounds (factory)
		Targa: 2,730 pounds
		Turbo: 2,867 pounds (factory)
0–60 mph:		911S: 6.3 seconds (*Road & Track*)
		Turbo: 4.9 seconds (*Road & Track*) add 1.5 seconds with Special Order
Maximum Speed:		911SC: 126 mph (*Road & Track*)
		Turbo: 165 mph (*Road & Track*) subtract 11 mph with Special Order
Brakes:		911SC: vented disc brakes
		Turbo: 917-style cross-drilled vented discs with four-piston alloy calipers
Steering:		ZF rack-and-pinion
Suspension:	Front:	MacPherson struts with telescoping shock absorbers, lower wishbones, longitudinal torsion bars, antisway bar
	Rear:	Telescoping shock absorbers, two-piece trailing arms, transverse torsion bars, antisway bar
Tires:	Front:	911SC: 185/70HR15
		optional: 205/55VR16
		Turbo: 205/55VR16
	Rear:	911S: 215/60VR15
		optional: 225/50VR16
		Turbo: 225/50VR16
Tire air pressure:	Front:	28 psi; Rear: 34 psi
		Turbo: Front: 28 psi; Rear: 42 psi
Transmission(s):		911SC: 915/61 5-speed U.S.
		911SC: 925/17 Special Order M09
		Sportomatic U.S.
		Turbo: 930/34 4-speed

Wheel:	Front:	911S: 6.0Jx15 ATS Cookie Cutter
		Optional: 6.0Jx16 Fuchs alloy
		Turbo: 7.0Jx16 Fuchs alloy
	Rear:	911S: 7.0Jx15 ATS Cookie Cutter
		Optional: 7.0Jx16 Fuchs alloy
		Turbo: 8.0Jx16 Fuchs alloy

What they said at the time—Porsche for 1978

Car and Driver, March 1978

"The new 911SC is faithful to tradition. It is the fastest normally aspirated Porsche, 0–60, that we have ever driven. It does the quarter-mile in 14.8 seconds at 94 miles per hour, and the factory (which is usually conservative in these things) rates its top speed at 136. While all this *sturm und drang* is slowly being fed through your mental computer, you must also come to grips with the information that it gets 15 miles per gallon in the EPA city cycle and 27 in the highway test. It is thus terribly fast and surprisingly economical. A remarkable blending of opposing virtues."

Parts List for 1978 911s

These are items most commonly replaced during regular maintenance and routine daily operation. Prices quoted are for new factory parts at list price, not including installation labor. NLA means factory parts no longer available, so prices quoted are from aftermarket suppliers.

Engine:

1. Oil filter $18.52
2. Alternator belt $12.72
3. Starter $433.15
4. Alternator $1,546.62
5. Muffler (50 states) $1,347.87
   (Turbo) $1,839.06
6. Clutch disc $440.19
   (Turbo) $890.29

Body:

7. Front bumper $1,190.93
8. Left front fender $1,564.86
9. Right rear
   quarter panel $1,977.49
10. Front deck lid $1,865.68
11. Front deck lid struts .. $43.22 each
12. Rear deck lid struts ... $32.88 each
13. Porsche badge,
    front deck lid $149.76
14. Taillight housing
    and lens $822.08
15. Windshield $1,011.43
16. Windshield weather
    stripping $155.44

Interior:

17. Dashboard (NLA) $1,534.00
18. Shift knob $97.42
19. Interior carpet,
    complete (NLA) .... $400–$800

Chassis:

20. Front rotor $205.80
21. Brake pads,
    front set $95.04
    (Turbo) $64.88
22. Koni rear
    shock absorber $213.50
23. Front wheel $1,374.67
24. Rear wheel $1,295.97

Ratings

1978 models, manual transmission (49-state cars)

	911SC coupe	911 Turbo 3.3 coupe	911SC Targa
Acceleration	4	5	4
Comfort	4	5	4
Handling	4	3m	3.5
Parts	4	4	3b
Reliability	4	4	4

b - Targa roofs no longer available from Porsche.

m - Turbo oversteer, see text.

Ratings continued

1978 models, manual transmission (California cars)

	911SC coupe	911 Turbo 3.3 coupe	911SC Targa
Acceleration	3.5	4.5	3.5
Comfort	4	5	4
Handling	4	3m	3.5
Parts	4h	4h	3bh
Reliability	4j	4j	4j

b - Targa roofs no longer available from Porsche.

h - Smog control equipment may be needed for licensing.

j - Air pump may provide better performance, see text.

m - Turbo oversteer, see text.

1978 models, Sportomatic transmission (49-state cars)

	911SC coupe	911SC Targa
Acceleration	3n	3n
Comfort	4	4
Handling	3.5	3
Parts	3d	2.5bd
Reliability	3en	3en

b - Targa roofs no longer available from Porsche.

d - Sportomatic special order only in 1978; parts, gaskets extremely hard to find, costly to buy.

e - Sportomatic repair difficult, expensive, may fail quickly. See text.

n - Sportomatic now three-speed.

1978 models, Sportomatic transmission (California cars)

	911SC coupe	911SC Targa
Acceleration	2.5n	2.5n
Comfort	4	4
Handling	3.5	3
Parts	2dh	2bdh
Reliability	2.5ejn	2.5ejn

b - Targa roofs no longer available from Porsche.

d - Sportomatic special order only in 1978; parts, gaskets extremely hard to find, costly to buy.

e - Sportomatic repair difficult, expensive, may fail quickly. See text.

h - Smog control equipment may be needed for licensing.

j - Air pump may provide better performance, see text.

n - Sportomatic now three speeds, see text.

1978 Garage Watch
Problems with (and improvements to) Porsche 911 models.

Improvements and solutions were everywhere, and some caused new problems. The previous 911S and Carrera got combined and renamed the 911SC, for Super Carrera. Porsche installed breakerless ignition on the SC. All Porsches worldwide now came with a redesigned power brake system.

Idle problem appears. Porsche reversed direction of distributor spin due to breakerless ignition, so what looks like spark retard now can only advance timing. There is no vacuum retard on distributor. To stabilize the idle, usual practice is to replace with later distributor or vacuum unit.

Beware of gray-market cars; catalytic converters without heat shields; exhaust gas recirculation systems; charcoal air filter canisters hung by hose clamps; electrical short cuts. Perhaps most dangerous are improper side door impact beams. (One converter installed wooden broom handles painted black to look like steel rods.) See Chapter 3 for specific information.

Heater blowers fail, as from 1975; front fresh air blowers continue to fail, as from 1969.

Rust behind permanent side reflectors, and in headlight buckets.

Brake master cylinders leak. This problem disappeared with the introduction of power brakes and its resulting complete redesign of the brake system for model year 1977, but all master cylinders can leak after about 10 years, as from 1965.

Aftermarket air conditioning compressors; factory air conditioning is through the dash and is poorly ducted. Factory standard on 930 Turbo—also automatic heat control on Turbo. (NOTE: Air conditioning systems, whether aftermarket or Porsche factory produced, are mediocre at best.)

Clutch modifications to handle higher torque, smoother starts. SC model introduced and torque damper (rubber center) of clutch disc disintegrates. All replaced by this time.

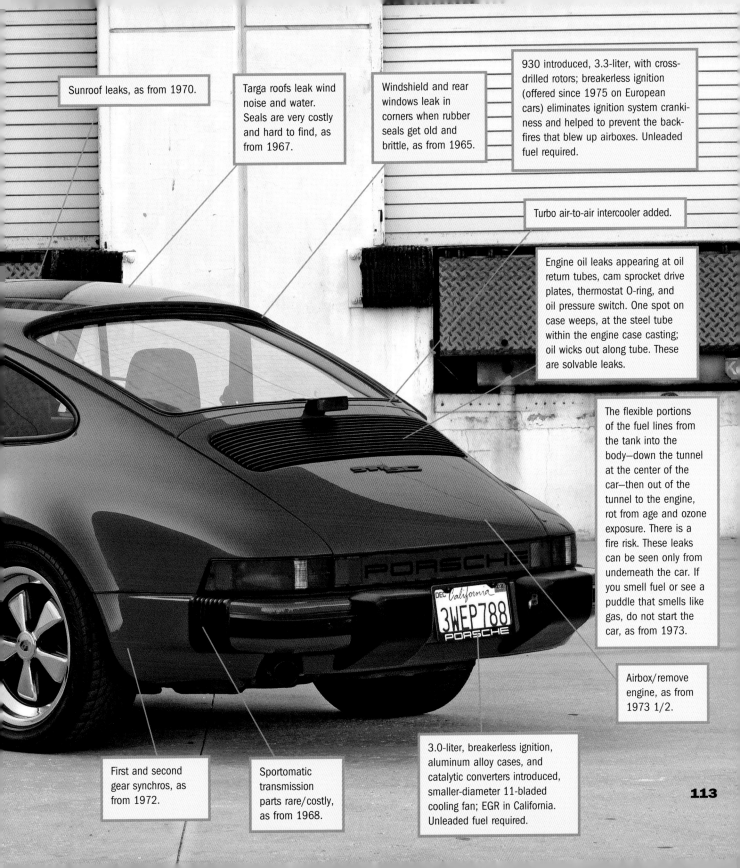

Sunroof leaks, as from 1970.

Targa roofs leak wind noise and water. Seals are very costly and hard to find, as from 1967.

Windshield and rear windows leak in corners when rubber seals get old and brittle, as from 1965.

930 introduced, 3.3-liter, with cross-drilled rotors; breakerless ignition (offered since 1975 on European cars) eliminates ignition system crankiness and helped to prevent the backfires that blew up airboxes. Unleaded fuel required.

Turbo air-to-air intercooler added.

Engine oil leaks appearing at oil return tubes, cam sprocket drive plates, thermostat O-ring, and oil pressure switch. One spot on case weeps, at the steel tube within the engine case casting; oil wicks out along tube. These are solvable leaks.

The flexible portions of the fuel lines from the tank into the body—down the tunnel at the center of the car—then out of the tunnel to the engine, rot from age and ozone exposure. There is a fire risk. These leaks can be seen only from underneath the car. If you smell fuel or see a puddle that smells like gas, do not start the car, as from 1973.

Airbox/remove engine, as from 1973 1/2.

First and second gear synchros, as from 1972.

Sportomatic transmission parts rare/costly, as from 1968.

3.0-liter, breakerless ignition, aluminum alloy cases, and catalytic converters introduced, smaller-diameter 11-bladed cooling fan; EGR in California. Unleaded fuel required.

Some critics complained that Porsche's 1978 SC was just in the middle of the S and the Carrera in trim and performance. For those critics, 1979 was even more frustrating as Porsche changed nothing on either its normally aspirated SC models or on the Turbo. Yet, with the change in 1978 to aluminum crankcases, the vacuum-boosted power brake system, and the breakerless capacitive discharge ignition (CDI) system, these cars had become more driveable and more economical to operate. In the passenger compartment, the SC models got the same automatic temperature control feature that had been standard on the Turbo models since their introduction in model year 1976.

The new 3.0-liter SC model, introduced in model year 1978, simplified Porsche's product line-up by installing the U.S.-required air pump on European models. There still were differences, however, between U.S. models and those for the rest of the world.

Porsche had improved engine cooling with the 1978 SC. This was especially important on U.S.-market cars. These engines generated much more engine heat in burning uncombusted gasses while being limited far below European autobahn-autoroute-autostrada speeds. At the same time, Porsche had needed to increase fan and belt speed in 1976 in order to run the alternator faster to meet the increased electrical requirements. In order to avoid overcooling the engine, Porsche changed from the original 11-bladed fan that ran at 1.3-times engine speed, to a 5-bladed fan that spun at 1.8 times engine speed. This worked great for the alternator but it proved inadequate for cooling U.S. emissions-controlled engines, especially on U.S. highways where speed was limited to 55 miles per hour (and engines loafed along at 2,500 rpm). In addition, some owners complained the fewer-bladed fan with more space between its vanes was noisier. So for model year

1978, Porsche went back to the 11-blade fan. However, to keep adequate cooling, it also kept the 1.8-to-1-fan ratio. This would remain through model year 1979, when Porsche would again revise this for 1980.

The three-speed Sportomatic was in its final year. For 1978 and 1979, Porsche made the transmission a special order option only (M09) and removed it from its regular catalogs in preparation for phasing it out entirely. (It would be gone by model year 1980.)

1979 Specifications "L" Series

Body Designation:		911SC, Turbo
Price:		911SC: $20,775
		Targa: add $1,225
		Turbo: $42,520
Engine Displacement and Type:		911SC (49-states)
		Typ 930/04; 2,994 cc (182.6 cid) SOHC, Bosch K-Jetronic fuel injection (CIS), air pump, thermal reactors, Catalytic converters
		911SC (California)
		Typ 930/06; 2,994 cc (182.6 cid) SOHC, Bosch K-Jetronic fuel injection (CIS), air pump, thermal reactors, EGR, Catalytic converters
		Turbo (49 states)
		Typ 930/61; 3,299 cc (201.2 cid), air-to-air intercooler, Bosch CIS, air pump, thermal reactors, EGR
		Turbo (California)
		Typ 930/63; 3,299 cc (201.2 cid), air-to-air intercooler, Bosch CIS air pump, thermal reactors, EGR, vacuum control ignition retard
Maximum Horsepower @ rpm:		911SC: 49 states and California:
		172 SAE @ 5,500 rpm
		Turbo: 253 SAE @ 5,500 rpm
Maximum Torque @ rpm:		911SC: 49-states and California
		189 ft-lb @ 4,000 rpm
		Turbo: 282 @ 4,000 rpm
Weight:		911SC: 2,560 pounds
		Turbo: 3,040 pounds
0–60 mph:		911SC: 6.0
		Turbo: 5.4 seconds (*Car and Driver*) add 1.5 seconds with Special Order Sportomatic
Maximum Speed:		911SC: 136 mph
		Turbo: 153 mph (*Car and Driver*) subtract 11 mph with Special Order Sportomatic
Brakes:		911SC: vented disc brakes
		Turbo: 917-style cross-drilled vented discs with four-piston alloy calipers
Steering:		ZF rack-and-pinion
Suspension:	Front:	MacPherson struts with telescoping shock absorbers, wishbones, torsion bars, antisway bar
	Rear:	telescoping shock absorbers, two-piece trailing arms, transverse torsion bars, antisway bar
Tires:	Front:	911SC: 185/70HR15
		optional: 205/55VR16
		Turbo: 205/55VR16
	Rear:	911SC: 215/60VR15 [Qu: SC?]
		Optional: 225/50VR16
		Turbo: 225/50VR16
Tire air pressure:	Front:	28 psi; Rear: 34 psi
		Turbo: Front: 28 psi; Rear: 42 psi
Transmission(s):		911SC: 915/63 5-speed U.S.
		911SC: 925/17 Special Order M09
		Sportomatic U.S.
		Turbo: 930/4 4-speed
Wheels	Front:	911SC: 6.0Jx15 ATS Cookie Cutter
		Optional: 6.0Jx16 Fuchs alloy
		Turbo: 7.0Jx16 Fuchs alloy
	Rear:	911S: 7.0Jx15 ATS Cookie Cutter
		Optional: 7.0Jx16 Fuchs alloy
		Turbo: 8.0Jx16 Fuchs alloy

What they said at the time—Porsche for 1979

Car and Driver, August 1979

"Porsche's continuing policy of constant improvement has added further intricacy to an already complex mechanical package, and last year's intercooler and running-gear changes have done their part to kick the duffs of the competition harder than ever before."

Parts List for 1979 911s

These are items most commonly replaced during regular maintenance and routine daily operation. Prices quoted are for new factory parts at list price, not including installation labor. NLA means factory parts no longer available, so prices quoted are from aftermarket suppliers.

Engine:

1. Oil filter.................... $18.52
2. Alternator belt........... $12.72
3. Starter...................... $433.15
4. Alternator................. $1,546.62
5. Muffler (50 states).... $1,347.87
   (Turbo) $1,839.06
6. Clutch disc............... $440.19
   (Turbo) $890.29

Body:

7. Front bumper............. $1,190.93
8. Left front fender........ $1,564.86
9. Right rear
   quarter panel............. $1,977.49
10. Front deck lid........... $1,865.68
11. Front deck lid struts.. $43.22 each
12. Rear deck lid struts... $32.88 each
13. Porsche badge,
    front deck lid............. $149.76
14. Taillight housing
    and lens $822.08
15. Windshield................. $1,011.43
16. Windshield weather
    stripping................... $155.44

Interior:

17. Dashboard (NLA)....... $1,534.00
18. Shift knob $97.42
19. Interior carpet,
    complete (NLA) .... $400–$800

Chassis:

20. Front rotor................. $205.80
21. Brake pads,
    front set $95.04
    (Turbo) $64.88
22. Koni rear
    shock absorber $213.50
23. Front wheel $1,374.67
24. Rear wheel $1,295.97

Ratings

1979 models, manual transmission (49-state cars)

	911SC coupe	911 Turbo 3.3 coupe	911SC Targa
Acceleration	4.5	5	4.5
Comfort	4	5	4
Handling	4	3m	3.5
Parts	4	4	3b
Reliability	4	4	4

b - Targa roofs no longer available from Porsche.
m - Turbo oversteer, see text.

1979 models, manual transmission (California cars)

	911SC coupe	911 Turbo 3.3 coupe	911SC Targa
Acceleration	4	4.5	4
Comfort	4	5	4
Handling	4	3m	3.5
Parts	4h	4h	3bh
Reliability	4j	4j	4j

b - Targa roofs no longer available from Porsche.
h - Smog control equipment may be needed for licensing.
j - Air pump may provide better performance, see text.
m - Turbo oversteer, see text.

Ratings continued

1979 models, Sportomatic transmission (49-state cars)

	911SC coupe	911SC Targa
Acceleration	3.5n	3.5n
Comfort	4	4
Handling	3.5	3
Parts	3d	2.5bd
Reliability	3en	3en

b - Targa roofs no longer available from Porsche.

d - Sportomatic special order, last year offered in 1979; parts, gaskets extremely hard to find, costly to buy.

e - Sportomatic repair difficult, expensive, may fail quickly. See text.

n - Sportomatic now three speeds, see text.

1979 models, Sportomatic transmission (California cars)

	911SC coupe	911SC Targa
Acceleration	3n	3n
Comfort	4	4
Handling	3.5	3
Parts	2dh	2bdh
Reliability	2.5ejn	2.5ejn

b - Targa roofs no longer available from Porsche.

d - Sportomatic special order only, last year available in 1979; parts, gaskets extremely hard to find, costly to buy.

e - Sportomatic repair difficult, expensive, may fail quickly. See text.

h - Smog control equipment may be needed for licensing.

j - Air pump may provide better performance, see text.

n - Sportomatic now three speeds, see text.

1979 Garage Watch
Problems with (and improvements to) Porsche 911 models.

Both model years 1978 and 1979 share many things in common, both good and bad. However, 1979 marked the end of smog pumps and Sportomatic transmissions.

Gray-market shortcuts, illegal, dishonest, unsafe, as from 1978. See Chapter 3 for gray-market information.

Last production year for U.S.-legal Turbos till 1986.

Sunroof leaks, as from 1970.

Windshield and rear windows leak in corners when rubber seals get old and brittle, as from 1965.

Rust behind permanent side reflectors and in headlight buckets. Once Porsche removed the glass over the headlight lens, this allowed water to seep behind the beam. Porsche cut a drain hole but this often filled from the bottom with road dirt, trapping moisture inside, as from 1968.

Aftermarket air conditioning compressors, and factory optional units, with problems, as from 1976.

Brake master cylinders leak. This problem disappeared with the introduction of power brakes and its resulting complete redesign of the brake system for model year 1977, but all master cylinders can leak after about 10 years, as from 1965.

The flexible portions of the fuel lines from the tank into the body—down the tunnel at the center of the car—then out of the tunnel to the engine, rot from age and ozone exposure. There is a fire risk. These leaks can be seen only from underneath the car. If you smell fuel or see a puddle that smells like gas, do not start the car, as from 1973.

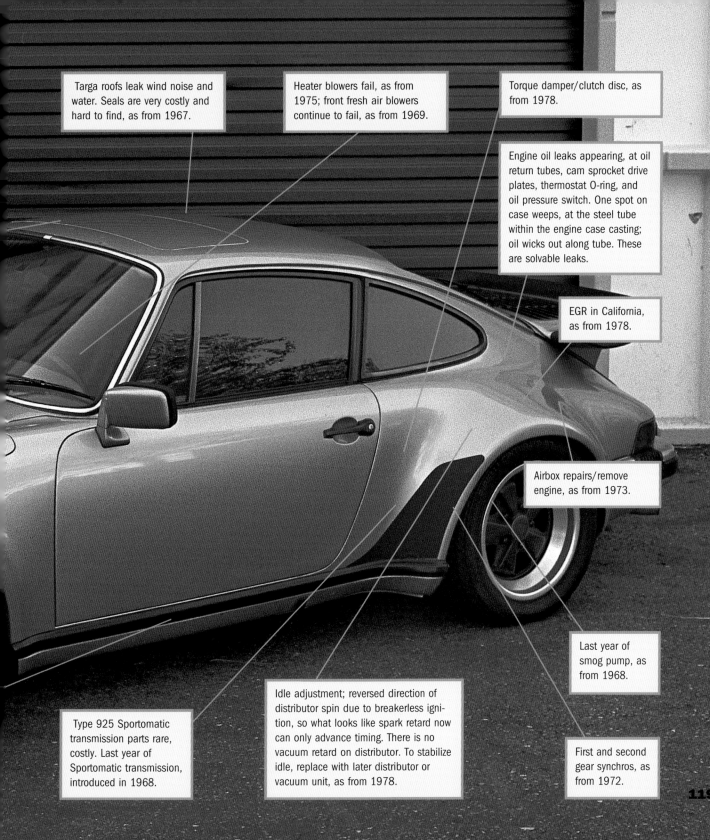

Targa roofs leak wind noise and water. Seals are very costly and hard to find, as from 1967.

Heater blowers fail, as from 1975; front fresh air blowers continue to fail, as from 1969.

Torque damper/clutch disc, as from 1978.

Engine oil leaks appearing, at oil return tubes, cam sprocket drive plates, thermostat O-ring, and oil pressure switch. One spot on case weeps, at the steel tube within the engine case casting; oil wicks out along tube. These are solvable leaks.

EGR in California, as from 1978.

Airbox repairs/remove engine, as from 1973.

Last year of smog pump, as from 1968.

Type 925 Sportomatic transmission parts rare, costly. Last year of Sportomatic transmission, introduced in 1968.

Idle adjustment; reversed direction of distributor spin due to breakerless ignition, so what looks like spark retard now can only advance timing. There is no vacuum retard on distributor. To stabilize idle, replace with later distributor or vacuum unit, as from 1978.

First and second gear synchros, as from 1972.

119

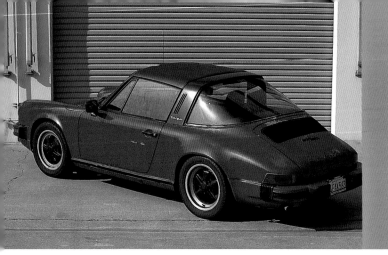

1980

Chapter 20
The New "A" Program—Rumors of the 911's Demise Are Exaggerated

U.S. emissions and safety standards finally got to Porsche. It could no longer rein in its exuberant Turbo to meet carbon monoxide, nitrogen oxide, and hydrocarbon levels that the U.S. Environmental Protection Agency and the California Air Resources Board had mandated for cars sold in the United States and in California. While it continued to offer the Turbo to customers in Europe, it no longer sold it in the United States or Canada, thereby launching a new enterprise, the gray-market car. Porsche had no trouble making the 1980 model year SC 50-state compliant and there no longer was any difference between the two cars.

The newest addition to Porsche's fight against noxious emissions was the three-way catalytic converter, equipped with an oxygen-sensor-operated frequency valve set in the exhaust manifold. Called the *lambda sonde* system, this measured the oxygen in the exhaust gasses and then adjusted the fuel-air mixture injected into the engine to ensure complete combustion of waste products in the catalytic converters. *Lambda*, a Greek letter, is also a technical symbol for the perfect combustion goal of CO_2 with neither excess oxygen nor carbon molecules remaining.

Porsche had improved engine cooling with the 1978 SC with an 11-blade fan and a 1.8-to-1-fan ratio. This remained through model year 1979 but for model year 1980, Porsche slowed the rotation speed down to 1.67 times engine speed. It remains that to this day.

After nearly negligible sales of the M09 special order three-speed Sportomatic in 1978 and 1979, Porsche discontinued the transmission and it was no longer available. These semiautomatic torque-converter transmissions had been devised to appeal to first-time U.S. customers who were uncomfortable with a clutch or who faced heavy traffic conditions in their daily commutes. Parts for these transmissions are increasingly hard to find, and the seals needed to complete an overhaul are almost nonexistent. Many owners choose instead to replace the Sportomatic with the Typ 915 four or five-speed manual transmissions when their semiauto fails.

Porsche made air conditioning and power window lifts standard equipment on U.S. cars, and it completed the stealth trim program begun in 1973 by blackening front air intakes. For model year 1980, Porsche trimmed its windows in black. With the exception of the alloy wheels, there was nothing on a 911 that distracted the viewer from the simple lines of the car's body form.

To emphasize that body form, and because Porsche could no longer offer U.S. customers its Turbo, it made available a limited-edition M439 "Weissach" model. Porsche fitted 6-inch front alloy rims and 7-inch rears, mounted with Pirelli Cinturato CN36 185/70R15 tires front and 215/60s on the rear. The car was available only in Black Metallic or in Platinum Metallic paint, with a Doric Gray full leather interior trimmed with Burgundy piping. Porsche also fitted the whale tail, front spoiler with rubber rub strip, power sunroof and power antenna, and extra stereo speakers in the doors and rear parcel shelf. Porsche produced 408 of these.

Finally, at the end of calendar year 1980, Managing Director (and Technical Director) Ernst Fuhrmann resigned, one year before his contract was to end. He had brought the racing Turbo to the streets and had supervised Porsche's war against automotive exhaust emissions. He also had introduced Porsche's front-engined water-cooled cars, the 924 and 928, and he was an enthusiastic supporter of these models, even while the Porsche and Piëch families retained their faith in the long-successful (and profitable) 911s. Ferry Porsche believed there was much development and life still in the venerable air-cooled cars. His nephew Ferdinand Piëch, who had become development director at Audi, believed all-wheel drive offered technology valuable to the 911. What's more, both families had argued for an open 911, which Fuhrmann had rejected, calling the 911 an obsolete design.

Porsche's board replaced Fuhrmann with German-born engineer Peter Schutz. He understood his bosses, and immediately launched new development programs for the 911 after he started in January 1981. On the Porsche stand at the October 1981 Frankfurt Motor Show, Porsche and Schutz showed the world an all-wheel-drive Turbo Cabriolet. Rumors of the demise of the 911 model promptly disappeared.

1980 Specifications "A" Series

Body Designation:		911SC, Turbo
Price:		911SC: $27,700
		Targa: add $1,450
Engine Displacement and Type:		911SC (50 states)
		Typ 930/04; 2,994 cc (182.6 cid) SOHC, Bosch K-Jetronic fuel injection (CIS), oxygen sensor (*lambda sonde*) with three-way catalytic converters
Maximum Horsepower @ rpm:		911SC: 50 states:
		172 SAE @ 5,500 rpm
Maximum Torque @ rpm:		911SC: 50 states
		189 ft-lb @ 4,000 rpm
Weight:		911SC: 2,805 pounds
0–62 mph:		911SC: 6.0 seconds (*Car and Driver*)
Maximum Speed:		911SC: 130 mph (*Car and Driver*)
Brakes:		911SC: vented disc brakes
Steering:		ZF rack-and-pinion
Suspension:	Front:	MacPherson struts with telescoping shock absorbers, lower wishbones, longitudinal torsion bars, antisway bar
	Rear:	telescoping shock absorbers, two-piece trailing arms, transverse torsion bars, antisway bar
Tires:	Front:	911SC: 215/60VR16
	Rear:	911SC: 225/50VR16
Tire air pressure:	Front:	28 psi; Rear: 34 psi
Transmission(s):		911SC: 915/63 5-speed U.S.
Wheels:	Front:	911SC: 6.0Jx16 Fuchs alloy
	Rear:	911SC: 7.0Jx16 Fuchs alloy

What they said at the time–Porsche for 1980

Car and Driver, August 1980

"Riding shotgun in a raging, spitting, hair-trigger 911SC is just about the most awful and impossible way there is to remain passive. This 911SC, this *thing*, could draw screams from the pope. Confronted with streaming, onrushing impedimenta, even the most stolid passengers collapse into beyond-the-pale-of-reason seizures of fright. 'This is like a roller coaster,' they whimper. 'I hate roller coasters.' But the 911SC is addictive."

Parts List for 1980 911s

These are items most commonly replaced during regular maintenance and routine daily operation. Prices quoted are for new factory parts at list price, not including installation labor. NLA means factory parts no longer available, so prices quoted are from aftermarket suppliers.

Engine:

1. Oil filter..................... $15.82
2. Alternator belt.......... $12.72
3. Starter....................... $433.15
4. Alternator................. $1,564.62
5. Muffler (50 states).... $1,347.87
6. Clutch disc............... $440.19

Body:

7. Front bumper............. $1,564.86
8. Left front fender........ $1,190.93
9. Right rear quarter panel............. $1,977.49
10. Front deck lid........... $1,865.68
11. Front deck lid struts.. $43.22 each
12. Rear deck lid struts... $32.88 each
13. Porsche badge, front deck lid............. $149.76
14. Taillight housing and lens.................... $822.08
15. Windshield................. $1,011.43
16. Windshield weather stripping $155.44

Interior:

17. Dashboard (NLA)....... $1,534.00
18. Shift knob $97.42
19. Interior carpet, complete (NLA) .... $400–$800

Chassis:

20. Front rotor................. $205.80
21. Brake pads, front set.................... $95.04
22. Koni rear shock absorber.......... $213.50
23. Front wheel $1,374.67
24. Rear wheel $1,295.97

Ratings

1980 models, manual transmission (only), 50-state USA

	911SC coupe	911SC Targa	911 Turbo
Acceleration	4	4	5p
Comfort	4.5	4.5	5p
Handling	4.5	4	3.5mp
Parts	4	3.5b	1p
Reliability	4	4	3p

b - Targa roofs no longer available from Porsche.

m - Turbo oversteer, see text.

p - Turbo not sold in U.S. Gray-market imports only. See text.

1980 Garage Watch
Problems with (and improvements to) Porsche 911 models.

All U.S. models now share common emissions control devices, known as 50-state cars. However, in response to perceived continued petroleum shortages and steady lobbying from Congress, speedometers on Porsches and other makes read only to 85 miles per hour.

Airbox repairs require engine removal, as from 1973 1/2.

Fuel injection calibration/idle adjustment, difficult to adjust on U.S. cars, through 1983. A plug blocked access to fuel mixture adjustment screw. By now all have been remedied.

Lambda sonde system for fuel injection.

Improved timing chain tensioner and idler arm.

Last year of rubber damper clutches, as from 1978.

First and second gear synchros, as from 1972.

The flexible portions of the fuel lines from the tank into the body—down the tunnel at the center of the car—then out of the tunnel to the engine, rot from age and ozone exposure. There is a fire risk. These leaks can be seen only from underneath the car. If you smell fuel or see a puddle that smells like gas, do not start the car, as from 1973.

Brake master cylinders leak. This problem disappeared with the introduction of power brakes and its resulting complete redesign of the brake system for model year 1977, but all master cylinders can leak after about 10 years, as from 1965.

Oxygen sensor with electronic control under passenger seat, replaced California-only system as from 1979.

Sunroof leaks, as from 1970.

Targa roofs leak wind noise and water. Seals are very costly and hard to find, as from 1967.

Heater blowers fail, as from 1975; front fresh air blowers continue to fail, as from 1969.

Gray-market shortcuts, as from 1978; no U.S.-legal Turbo as of 1980 model year. See Chapter 3.

Rust behind permanent side reflectors, and in headlight buckets. Once Porsche removed the glass over the headlight lens, this allowed water to seep behind the beam. Porsche cut a drain hole but this often filled from the bottom with road dirt, trapping moisture inside. Factory got zinc undercoating mix correct to hold paint, allowing improved metallic options, as from 1968.

First year of factory standard air conditioning on U.S. delivery cars; however, the same considerations should apply, with leaking Freon lines, condenser placement, and compressor mounts, as from 1976.

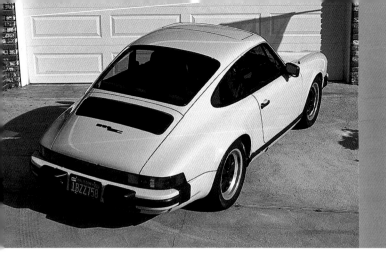

1981

Chapter 21
The New "B" Program—
Rumors of the 911's
Demise Are Still
Exaggerated

U.S. emissions and safety standards kept the Turbo out of the United States again in 1981 (though it was now offered in Canada). As a result, a growing aftermarket, or more accurately, outside enterprise, grew up to "federalize" non-U.S.-legal high-performance models. These "gray-market" converters, some very good, honest, and responsible with solid engineering knowledge, and some shady charlatans who knew only their bank balance, hung out their shingles as importers. Generally it is wise to stay away from gray-market cars. There are many more questions than answers, and none of the converters are still in business to reply. Some met federal side impact safety standards by painting a broom handle black and screwing it in between the door panels. It looked like steel but it wouldn't hold a magnet, nor would it save your life.

To identify non-U.S. models, look at the 17-digit VIN number. U.S. models will read WPOAA with 12 numbers and letters after the double A. If you see WPOZZZ (or the Canadian WPOJ), it is important to understand that you cannot get financing to purchase this car and you may not be able to insure it. If you are looking only for a "track car," a vehicle for weekend outings with Porsche Owners Club members on closed tracks, this would be an option for you. But for on-road use, you may find it's more trouble than it's worth.

The gray market first surfaced in 1977 as European Turbos and Carreras provided much greater horsepower than the U.S. versions and some U.S. buyers had to have this extra performance. This business flourished from early calendar year 1980 through the late 1980s, even though Porsche began importing its Turbo model again in model year 1986.

For those customers in Canada and the rest of the world, a new Turbo option first appeared in model year 1981 from Rolf Sprenger's *Sonderwunsch*, or Special Wishes, department. Eventually designated the M505 for U.S. and M506 for rest-of-world customers, it was the *flachtbau*, or flat nose, body modification. This took the flattened front fenders of Porsche's racing Typ

935 and mounted them on road cars. This would remain a special order option until model year 1987, a year after the Turbo would return to U.S. customers. At that time, it would become a regular production option.

To further diminish the risk of an airbox destroying backfire, Porsche revised the Bosch K-Jetronic fuel injection system by adding a cold-start mixture injector inside the airbox rather than injecting the mixture into the further distant air distributor housing. The mix often had gone too lean by the time it reached the airbox and this, Porsche engineers determined, was the cause of most airbox failures. (For the record, there are exceptions to every rule; there are still engines that will backfire, even with every one of these modifications and improvements in place. A prudent buyer would carefully scrutinize service records looking for a second or third airbox replacement. If that situation appears, this would be a car to avoid. This is a job requiring the shop to remove the engine from the car.)

Porsche upgraded the earlier braided fuel lines to seamless stainless steel lines contoured to the engine and engine compartment. Limited stretches of braided line served as connections between the fixed stainless lines and the engine to allow for vibration.

Between the *lambda sonde* and other changes to the K-Jetronic fuel injection system, and by raising the compression from 8.6:1 up to 9.3:1 in the United States, Porsche greatly improved fuel economy, by nearly 21 percent (to 29 miles per gallon highway/24 urban EPA rating) for the 1980 model year. (This came about also because of the introduction of premium grade unleaded fuels, required for the higher combustion engines.) While the 180 SAE net horsepower rating remained the same for the United States, increasing compression to 9.6:1 netted European, Canadian, and Japanese customers a large power increase, from 180 DIN to 204 DIN.

Porsche continued with rubber center clutches through model year 1980, returning to spring-centered clutch discs in all model year 1981 cars.

1981 Specifications "B" Series

Body Designation:		911SC
Price:		911SC: $28,365
		Targa: add $1,450
Engine Displacement and Type:		911SC (50 states) Typ 930/16; 2,994 cc (182.6 cid) SOHC, Bosch K-Jetronic fuel injection (CIS), Oxygen sensor (*lambda sonde*) with 3-way Catalytic converters
Maximum Horsepower @ rpm:		911SC: 50 states:
		172 SAE @ 5,500 rpm
Maximum Torque @ rpm:		911SC: 50 states
		189 ft-lb @ 4,000 rpm
Weight:		911SC: 2,560 pounds
0–60 mph:		911SC: 6.7 seconds
Maximum Speed:		911SC: 130 mph
Brakes:		911SC: vented disc brakes
Steering:		ZF rack-and-pinion
Suspension:	Front:	MacPherson struts with telescoping shock absorbers, lower wishbones, longitudinal torsion bars, antisway bar
	Rear:	telescoping shock absorbers, two-piece trailing arms, transverse torsion bars, antisway bar
Tires:	Front:	911SC: 205/60VR17
	Rear:	911SC: 225/55VR16
Tire air pressure:	Front:	28 psi; Rear: 34 psi
Transmission(s):		911SC: 915/63 5-speed U.S.
Wheels:	Front:	911SC: 6.0Jx16 Fuchs alloy
	Rear:	911SC: 7.0Jx16 Fuchs alloy

What they said at the time—Porsche for 1981

Car and Driver, October 1980

By David E. Davis

"This Porsche feels old, somehow. It feels as if it's finally coming to the end of its allotted life span. I'm no longer willing to put up with the harshness, or the way the front end hunts and nibbles when you power it through a favorite country corner."

Parts List for 1981 911s

These are items most commonly replaced during regular maintenance and routine daily operation. Prices quoted are for new factory parts at list price, not including installation labor. NLA means factory parts no longer available, so prices quoted are from aftermarket suppliers.

Engine:

1. Oil filter..................... $15.82
2. Alternator belt.......... $12.72
3. Starter....................... $389.63
4. Alternator................. $1,546.62
5. Muffler (50 states).... $1,347.87
6. Clutch disc............... $440.19

Body:

7. Front bumper............. $1,190.93
8. Left front fender........ $1,564.86
9. Right rear quarter panel............ $1,977.49
10. Front deck lid............ $1,865.68
11. Front deck lid struts.. $43.22 each
12. Rear deck lid struts... $32.88 each
13. Porsche badge, front deck lid............. $149.76
14. Taillight housing and lens.................... $822.08
15. Windshield................. $1,011.43
16. Windshield weather stripping.................... $155.44

Interior:

17. Dashboard (NLA)....... $1,534.00
18. Shift knob $97.42
19. Interior carpet, complete (NLA) .... $400–$800

Chassis:

20. Front rotor................. $205.80
21. Brake pads, front set $95.04
22. Koni rear shock absorber.......... $213.50
23. Front wheel $1,374.67
24. Rear wheel $1,295.97

Ratings

1981 models, manual transmission

	911SC coupe	911SC Targa	911 Turbo
Acceleration	4	4	5p
Comfort	4.5	4.5	5p
Handling	4.5	4	3.5mp
Parts	4	3.5b	1p
Reliability	4.5	4.5	3p

b - Targa roofs no longer available from Porsche.

m - Turbo oversteer, see text.

p - Turbo not sold in U.S. Gray-market imports only. See text.

1981 Garage Watch
Problems with (and improvements to) Porsche 911 models.

Few changes, several improvements, one a mixed blessing.

Windshield and rear windows leak in corners when rubber seals get old and brittle, as from 1965.

Rust behind permanent side reflectors, and in headlight buckets. Once Porsche removed the glass over the headlight lens, this allowed water to seep behind the beam. Porsche cut a drain hole but this often filled from the bottom with road dirt, trapping moisture inside, as from 1968.

Sunroof leaks, as from 1970.

Targa roofs leak wind noise and water. Seals are very costly and hard to find, as from 1967.

Gray-market horrors, as from 1978; some Canadian Turbos were snuck into U.S. with no thermal reactors. See Chapter 3.

Timing chain tensioner, as from 1968.

First and second gear synchros, 915 transmissions, as from 1972.

All fuel lines, injector lines now steel.

Improved airbox now has cold-start mixture distributor, still fails, requiring engine removal to replace the airbox.

Steel spring clutches replaced rubber torque-damper, as from 1978.

The flexible portions of the fuel lines from the tank into the body—down the tunnel at the center of the car—then out of the tunnel to the engine, rot from age and ozone exposure. There is a fire risk. These leaks can be seen only from underneath the car. If you smell fuel or see a puddle that smells like gas, do not start the car, as from 1973.

Brake master cylinders leak. This problem disappeared with the introduction of power brakes and its resulting complete redesign of the brake system for model year 1977, but all master cylinders can leak after about 10 years, as from 1965.

Heater blowers fail, as from 1975; front fresh air blowers continue to fail, as from 1969.

1982

Chapter 22
The New "C" Program—Rumors of the 911's Demise Continue Anyway

U.S. emissions and safety standards excluded American customers from the right to play with the Turbo for another year (though it remained available in Canada). The gray-market importers and "federalizers" continued to bring in cars for U.S. customers under a one-time exemption.

To further diminish the risk of airbox destroying backfire, Porsche revised the Bosch K-Jetronic fuel injection system by adding a cold-start injector spray. It also upgraded the earlier braided fuel lines to seamless stainless steel lines contoured to the engine and engine compartment. Limited stretches of braided line served as connections between the fixed stainless lines and the engine to allow for vibration.

In order to meet the growing electrical requirements, Porsche fitted a new 1,050-watt alternator that incorporated a voltage regulator on SC and Carrera models through model year 1989. Unfortunately, this new unit proved to have only a 40,000-mile life. As it approaches failure, it overcharges the battery. Overcharging makes the headlights surge, the air conditioner blower fan speed up and slow down, and the battery stink. If a noxious odor begins to pervade the passenger compartment, this likely means the alternator is overcharging the battery and beginning to fail. At failure, it can spike voltage, burning out other electric components and incurring costly repairs and replacements. Pay attention to any foul, unexplainable odor.

1982 Specifications "C" Series

Body Designation:		911SC
Price:		911SC: $29,165
		Targa: add $1,450
Engine Displacement and Type:		911SC (50 states) Typ 930/16; 2,994 cc (182.6 cid) SOHC, Bosch K-Jetronic fuel injection (CIS), oxygen sensor (*lambda sonde*) with 3-way Catalytic converters
Maximum Horsepower @ rpm:		911SC: 50 states:
		172SAE net @ 5,900 rpm
Maximum Torque @ rpm:		911SC: 50 states
		189 ft-lb @ 4,000 rpm
Weight:		911SC: 2,560 pounds
0–60 mph:		911SC: 6.0 seconds (*Car and Driver*)
Maximum Speed:		911SC: 130 mph (*Car and Driver*)
Brakes:		911SC: vented disc brakes
Steering:		ZF rack-and-pinion
Suspension:	Front:	MacPherson struts with telescoping shock absorbers, lower wishbones, longitudinal torsion bars, antisway bar
	Rear:	telescoping shock absorbers, two-piece trailing arms, transverse torsion bars; antisway bar
Tires:	Front:	911SC: 215/60HR16
	Rear:	911S: 225/50VR16
Tire air pressure:	Front:	28 psi; Rear: 34 psi
Transmission(s):		911SC: 915/63 5-speed U.S.
Wheels:	Front:	911S: 6.0Jx16 Fuchs alloy
	Rear:	911S: 7.0Jx16 Fuchs alloy

What they said at the time—Porsche for 1982

Sporting Cars, February 1982

"Generally, it is a well-tamed and stable platform, and long-striding fun to drive, for, in any gear, it always has that capability to erupt from traffic and charge away, almost matchless—except by another Porsche. Braking was always a joy (when necessary), progressive, accurate, and powerful with bags of feel. Ride is firm but stable and well-damped, comfortable at high speed though accompanied by poorly suppressed road noise from those broad tires."

Parts List for 1982 911s

These are items most commonly replaced during regular maintenance and routine daily operation. Prices quoted are for new factory parts at list price, not including installation labor. NLA means factory parts no longer available, so prices quoted are from aftermarket suppliers.

Engine:

1. Oil filter.................... $18.52
2. Alternator belt.......... $12.72
3. Starter....................... $433.15
4. Alternator................. $1,546.62
5. Muffler (50 states).... $1,347.87
6. Clutch disc............... $440.19

Body:

7. Front bumper............. $1,190.93
8. Left front fender........ $1,564.86
9. Right rear
    quarter panel............. $1,977.49
10. Front deck lid........... $1,865.68
11. Front deck lid struts.. $43.22 each
12. Rear deck lid struts... $323.88 each
13. Porsche badge,
     front deck lid............. $149.67
14. Taillight housing
     and lens.................... $822.08
15. Windshield................. $1,011.43
16. Windshield weather
     stripping.................... $155.44

Interior:

17. Dashboard (NLA)....... $1,534.00
18. Shift knob $97.42
19. Interior carpet,
     complete (NLA) .... $400–$800

Chassis:

20. Front rotor................ $205.80
21. Brake pads,
     front set $95.04
22. Koni rear
     shock absorber.......... $213.50
23. Front wheel $1,374.67
24. Rear wheel $1,295.97

Ratings

1982 models, manual transmission

	911SC coupe	911SC Targa	911 Turbo
Acceleration	4	4	5p
Comfort	4.5	4.5	5p
Handling	4.5	4	3.5mp
Parts	4	3.5b	1p
Reliability	3.5q	3.5q	2pq

b - Targa roofs no longer available from Porsche.

m - Turbo oversteer, see text.

p - Turbo not sold in U.S. Gray-market imports only. See text.

q - Alternator; see text.

1982 Garage Watch
Problems with (and improvements to) Porsche 911 models.

Gray-market horrors, as from 1978, no U.S.-legal Turbo, as from 1980. See Chapter 3.

Windshield and rear windows leak in corners when rubber seals get old and brittle, as from 1965.

Rust behind permanent side reflectors and in headlight buckets. Once Porsche removed the glass over the headlight lens, this allowed water to seep behind the beam. Porsche cut a drain hole but this often filled from the bottom with road dirt, trapping moisture inside, as from 1968.

Brake master cylinders leak, a problem continuing until model year 1990. The brakes would not fully release after being applied. This resulted from pedal bushings stiffening up or being swollen due to rust. The pistons stuck in the calipers. Brake hoses swelled shut, acting like a one-way valve. This problem disappeared with the introduction of power brakes and its resulting complete redesign of the brake system, but all master cylinders can leak after about 10 years, as from 1965.

New 1,050-watt alternator incorporates voltage regulator; however has average life of only 40,000 miles. Can lock on and spike voltage, burning out other electric components. As alternator begins to fail, it overcharges battery. Headlights will surge brighter and dimmer, heater or air conditioner fans will speed up or slow while on same setting. Overcharging makes the battery smell. If you notice a noxious odor, it is most likely the alternator overcharging the battery. This continues through till midyear 1989.

Heater blowers fail, as from 1975; front fresh air blowers continue to fail, as from 1969.

136

Targa roofs leak wind noise and water. Seals are very costly and hard to find, as from 1967.

Aftermarket cabriolets produced off Targa chassis, chassis flexed, leaking wind and water from windows, top.

Sunroof leaks, as from 1970.

Age problem appearing. The flexible portions of the fuel lines from the tank into the body—down the tunnel at the center of the car—then out of the tunnel to the engine, are beginning to rot from age and ozone exposure, causing fuel leaks at high pressure, especially in CIS cars, at about 75 psi. There is a fire risk. These leaks can be seen only from underneath the car. If you smell fuel or see a puddle that smells like gas, do not start the car, as from 1973.

First and second gear synchros, 915 transmissions, as from 1972.

Improved airbox still fails/remove engine, as from 1973 1/2 and 1981.

Timing chain tensioner, as from 1968.

Porsche began taking orders for its new SC model Cabriolet during the Geneva, Switzerland, auto show in March 1982. What rumors of impending 911 extinction the Frankfurt show hadn't silenced the previous October were finally set to rest as order books and Porsche bank accounts filled with deposits.

Buyers found a manual top that could be raised and lowered easily and quickly. Porsche fitted a zip-out rear plastic window that offered additional ventilation, but Porsche designers intended this feature as a way to more safely stow the plastic without scratching or creasing it. They also offered a zip-out glass window as well.

It was a success, as many first edition Porsches had been before it. When production totals were completed for model year 1983, Porsche had produced 4,277 of the new Cabrios but just 2,752 Targas (out of a total of 12,728 911SCs), proving that a good-sized enthusiast market had waited patiently for a fully open car.

U.S. emissions and safety standards yet again excluded U.S. customers from the right to purchase the Turbo (though it remained available in Canada, and Porsche produced 1,080 in model year 1983). The gray-market importers and "federalizers" continued to bring in cars for U.S. customers under a one-time exemption. These gray-market converters ranged in quality from very honest and responsible to very shady with little engineering knowledge. See Chapter 3 for a more full rundown on gray-market considerations.

Generally it is wise to avoid these non-U.S. models. To identify them, look at the 17-digit VIN number. U.S. models will read WPOAA with 12 numbers and letters after the double A. If you see WPOZZZ (or the prefix WPOJ that Porsche used for Canadian exports), this is a non-U.S. model. It is important to understand that you cannot get financing to purchase this car and you may not be able to insure it.

The gray market flourished from early calendar year 1980 through the late 1980s, even though Porsche began importing its Turbo model again in model year 1986.

Customers in Canada and the rest of the world could order a Turbo option from Rolf Sprenger's *Sonderwunsch*, or Special Wishes, department that first had appeared in model year 1981. Eventually designated the M505 for U.S. and M506 for rest-of-the-world customers, it was the *flachtbau* or flat nose body modification. This took the flattened front fenders of Porsche's racing Typ 935 and mounted them on road cars. This would remain a special order option until model year 1987 when, a year after the Turbo would return to U.S. customers, it became a regular production option.

To further diminish the risk of an airbox destroying backfire, Porsche revised the Bosch K-Jetronic fuel injection system by adding a cold-start injector spray inside the airbox itself, beginning in 1981. At the same time, it also upgraded the earlier braided fuel lines to seamless stainless steel lines contoured to the engine and engine compartment.

Porsche had introduced a most effective weapon against noxious emissions in model year 1980 with the three-way catalytic converter equipped with an oxygen sensor-operated frequency valve set in the exhaust manifold. Called the *lambda sonde* system, this measured the oxygen in the exhaust gasses and then adjusted the fuel-air mixture injected into the engine to ensure complete combustion of waste products in the catalytic converters. *Lambda*, a Greek letter, is also a technical symbol for the perfect combustion goal of CO_2 with neither excess oxygen nor carbon molecules remaining.

In order to meet the growing electrical requirements of electric windows, roofs, antennas, air conditioning, and security systems, Porsche fitted a new 1,050-watt alternator that incorporated a voltage regulator on SC and Carrera models through model year 1989. Unfortunately, this new unit proved to have only a 40,000-mile life. As it approaches failure, it overcharges the battery. Overcharging makes the headlights surge, the air conditioner blower fan speed up and slow down, and the battery stink. If a noxious odor begins to pervade the passenger compartment, this likely means the alternator is overcharging the battery and beginning to fail. At failure, it can spike voltage, burning out other electric components and incurring costly repairs and replacements. Pay attention to any unexplainable foul odor.

1983 Specifications "D" Series

Body Designation:		911SC
Price:		911SC coupe: $29,950
		911SC Targa: $31,450
		911SC Cabriolet: $34,450
Engine Displacement and Type:		911SC (50 states) Typ 930/16; 2,994 cc (182.6 cid) SOHC, Bosch K-Jetronic fuel injection (CIS), oxygen sensor (*lambda sonde*) with 3-way Catalytic converters
Maximum Horsepower @ rpm:		911SC: 50 states:
		172 SAE @ 5,500 rpm
Maximum Torque @ rpm:		911SC: 50 states
		189 ft-lb @ 4,200 rpm
Weight:		911SC: 2,560 pounds
		Cabrio: 2,750 pounds
0–60 mph:		911SC: 6.7 seconds
		Cabrio: 9.0 seconds (*Car and Driver*)
Maximum Speed:		911SC: 139 mph
		Cabrio: 136 mph (*Car and Driver*)
Brakes:		911SC: vented disc brakes
Steering:		ZF rack-and-pinion
Suspension:	**Front:**	MacPherson struts with telescoping shock absorbers, lower wishbones, longitudinal torsion bars, antisway bar
	Rear:	telescoping shock absorbers, two-piece trailing arms, transverse torsion bars, antisway bar
Tires:	**Front:**	911SC: 205/55VR16
	Rear:	911SC: 225/50VR16
Tire air pressure:	**Front:**	28 psi; Rear: 35 psi
Transmission(s):		911SC: 915/63 5-speed U.S.
Wheels:	**Front:**	911SC: 6.0Jx15 Fuchs alloy
	Rear:	911SC: 7.0Jx16 Fuchs alloy

139

What they said at the time–Porsche for 1983

Car and Driver, February 1983

"The 911SC convertible is the first major product for which [new Board Chairman Peter W.] Schutz can take the credit, so it is both a turning point for the firm and a portent of Porsches to come. As such, the Cabriolet is more phenomenon than fad: this is where Porsche shifts away from building pure engineering exercises and starts producing automobiles in response to legitimate market demands."

Parts List for 1983 911s

These are items most commonly replaced during regular maintenance and routine daily operation. Prices quoted are for new factory parts at list price, not including installation labor. NLA means factory parts no longer available, so prices quoted are from aftermarket suppliers.

Engine:

1. Oil filter..................... $18.52
2. Alternator belt.......... $12.72
3. Starter...................... $433.15
4. Alternator $1,546.62
5. Muffler (50 states).... $1,347.87
6. Clutch disc $440.19

Body:

7. Front bumper............. $1,190.93
8. Left front fender........ $1,564.86
9. Right rear quarter panel............. $1,977.49
10. Front deck lid $1,865.68
11. Front deck lid struts.. $43.22 each
12. Rear deck lid struts... $32.88 each
13. Porsche badge, front deck lid............ $149.67
14. Taillight housing and lens $822.08
15. Windshield................ $586.20
16. Windshield weather stripping................... $155.44

Interior:

17. Dashboard (NLA)....... $1,534.00
18. Shift knob $97.42
19. Interior carpet, complete (NLA) .... $400–$800

Chassis:

20. Front rotor................ $73.31
21. Brake pads, front set $95.04
22. Koni rear shock absorber.......... $213.50
23. Front wheel $1,374.67
24. Rear wheel $1,295.97

Ratings

1983 models, manual transmission

	911SC coupe	911SC Targa	911SC Cabrio	911 Turbo
Acceleration	4	3.5	3.0	5p
Comfort	4.5	4.5	3.5	5p
Handling	4.5	4	3.5	3mp
Parts	4	3b	3	1p
Reliability	3.5q	3.5q	3.5q	2pq

b - Targa roofs no longer available from Porsche.

m - Turbo oversteer, see text.

p - Turbo not sold in U.S. Gray-market imports only. See text.

q - Alternator, see text.

1983 Garage Watch
Problems with (and improvements to) Porsche 911 models.

Gray-market horrors as from 1978; still no U.S.-legal Turbo, as from 1980. See Chapter 3.

Factory manual top Cabriolet, very minor leaks, wind noises, manual lifting mechanisms rarely fail.

Rust behind permanent side reflectors and in headlight buckets.

Heater blowers fail, as from 1975; front fresh air blowers continue to fail, as from 1969.

Ride height now matches rest of world, lowered from 1975 elevation for 5 mile-per-hour safety bumpers.

Return of "real" speedometers, reading 160 miles per hour.

Windshield and rear windows leak in corners when rubber seals get old and brittle, as from 1965.

Age problem appearing. The flexible portions of the fuel lines from the tank into the body—down the tunnel at the center of the car—then out of the tunnel to the engine, are beginning to rot from age and ozone exposure, causing fuel leaks at high pressure, especially in CIS cars, at about 75 psi.

Targa roofs leak wind noise and water. Seals are very costly and hard to find, as from 1967.

Sunroof leaks, as from 1970.

Timing chain tensioner, as from 1968.

The company decided to revert to manual heat, air conditioning, and defroster controls with the Cabrio while the Targa and coupes retained fully automatic climate control operation.

New 1,050-watt alternator incorporates voltage regulator; however, has average life of only 40,000 miles.

Improved airbox still can fail, requiring engine removal to replace, as from 1973 1/2 and 1981.

Brake master cylinders leak, a problem continuing until model year 1990. The brakes would not fully release after being applied. This resulted from pedal bushings stiffening up or being swollen due to rust. The pistons stuck in the calipers. Brake hoses swelled shut, acting like a one-way valve. This problem disappeared with the introduction of power brakes.

First and second gear synchros, as from 1972.

1984

Chapter 24
The New "E" Program—The Good Gets Much, Much Better

The new Carrera 3.2 engine was made up of 80 percent new parts. European, Canadian, and Japanese customers got a car that reached 60 miles per hour in 5.3 seconds and topped out at 152, performance just a hair slower than the 1976 Turbo but without the thrill-giving Turbo lag.

To do this, Porsche replaced the continuous injection Bosch K-Jetronic fuel injection with an entirely new engine management system, Bosch's Digital Motor Electronics (DME). This operated the fuel injection and ignition timing by monitoring throttle setting, outside and engine temperatures, engine load, engine speed, and even battery voltage. Porsche engineers hid the DME "black box" under the driver's seat.

This new engine also addressed an age old (literally, from 1963) problem. With its overhead camshafts driven by chains, lubricating and maintaining proper tensions on the chains had been a consistent problem. Ernst Fuhrmann, seeing the end of the 911, had never authorized further improvements, but under Peter Schutz, Weissach engineers devised the Carrera chain tensioner and oiling system.

U.S. emissions and safety standards still kept U.S. customers from buying the Turbo (though it remained available in Canada). The gray-market importers and "federalizers" continued to bring in cars for U.S. customers under a one-time exemption. These gray-market converters ranged in quality from very honest and responsible to very shady with little engineering knowledge.

It is generally wise to avoid these non-U.S. models. To identify them, look at the 17-digit VIN number. U.S. models will read WPOAA with 12 numbers and letters after the double A. If you see WPOZZZ or WPOJ (Porsche's prefix for Canadian export cars), this is a non-U.S. model. It is important to understand that you cannot get financing to purchase this car and you may not be able to insure it.

The gray market flourished from early calendar year 1980 through the late 1980s, even though Porsche began importing its Turbo model again in model year 1986. It picked up steam particularly after England's Royal Auto Club officiated at a "supercar"

test in June 1984 and determined the Turbo was "The Fastest Accelerating Production Car in the World," mastering the standing-start kilometer (0.625 mile) at 135 miles per hour.

For customers in Canada and the rest of the world, a new Turbo option first appeared in model year 1981 from Rolf Sprenger's *Sonderwunsch*, or Special Wishes, department. Eventually designated the M505 for U.S. and M506 for rest-of-the-world customers, it was the *flachtbau* or flat nose body modification. This took the flattened front fenders of Porsche's racing Typ 935 and mounted them on road cars. This would remain a special order option until model year 1987, a year after the Turbo would return to U.S. customers.

Porsche upgraded the earlier braided fuel lines to seamless stainless steel lines contoured to the engine and engine compartment.

The new Cabrio body style surprised factory product planners with its sales through model year 1984, when it sold 3,103 copies, compared to the introductory model year 1983 with sales of 4,277. Targa sales, down in 1983 to 2,752, came back up in 1984, to 3,793 cars, out of total Carrera sales of 13,482. (In addition, rest-of-world Turbo sales dropped from 1,080 in 1983 to 881 in 1984. The burden was weighing heavily on Weissach engineers to make a U.S.-legal Turbo model, despite the fact that the new Carrera provided performance very close to what the original 1976 930S Turbo had offered.)

Cleverly sensing that as many owners wanted the look of the Turbo body as those seeking the additional performance, Porsche introduced option M491, the Turbo Look for Carrera coupes. This added the Turbo's front and rear flared fenders to the standard Carrera body, as well as its front and rear spoilers. Less visible but more effective, the package fitted Turbo front hubs, rear suspension arms and torsion bar tube, the entire Turbo brake system, and the Turbo's wheels and tires.

The new Carrera sold well. Porsche produced 14,309 of them in model year 1984, as well as nearly 26,800 of its new 944 models. More than 40 percent of Porsche's total production went to the United States.

1984 Specifications "E" Series

Body Designation:		911 Carrera
Price:		911 Carrera coupe: $31,950
		911 Carrera Targa: $33,450
		911 Carrera Cabriolet: $36,450
Engine Displacement and Type:		911 Carrera (50 states)Typ 930/16; 3,164 cc (192.6 cid) SOHC, Bosch K-Jetronic fuel injection (CIS), oxygen sensor (*lambda sonde*) with 3-way Catalytic converters
Maximum Horsepower @ rpm:		911 Carrera: 50 states: 200 SAE @ 5,900 rpm
Maximum Torque @ rpm:		911 Carrera: 50 states 185 ft-lb @ 4,800 rpm
Weight:		Coupe: 2,728 pounds
		Cabrio: 2,950 pounds
0–60 mph:		Coupe: 5.7 seconds (*Motor Trend*)
		Cabrio: 7.5 seconds (*Car and Driver*)
Maximum Speed:		Coupe: 146 mph (*Motor Trend*)
		Cabrio: 124 mph (*Car and Driver*)
Brakes:		911 Carrera: vented disc brakes
Steering:		ZF rack-and-pinion
Suspension:	Front:	MacPherson struts with telescoping shock absorbers, lower wishbones, longitudinal torsion bars, antisway bar
	Rear:	Telescoping shock absorbers, two-piece trailing arms, transverse torsion bars, antisway bar
Tires:	Front:	Carrera: 205/55ZR16
	Rear:	Carrera: 225/50ZR16
Tire air pressure:	Front:	28 psi; Rear: 35 psi
Transmission(s):		Carrera: 915/63 5-speed U.S.
Wheels:	Front:	Carrera: 6.0Jx16 Fuchs alloy optional: 7.0Jx16
	Rear:	Carrera: 7.0Jx15 Fuchs alloy optional: 8.0Jx16

What they said at the time–Porsche for 1984

Car and Driver, February 1984

"The Carrera replaces the 911SC, and it embodies all the same simultaneously outrageous and sensible qualities and more. The 911SC was fast, but the Carrera is a bullet. Firing from 0 to 60 miles per hour in 5.3 seconds, an improvement of more than a full second, 'shot out of a gun,' covers it. And, although the 911SC returned a resolute 16 miles per gallon on the EPA city cycle, the Carrera offers no less than a whopping 20 miles per gallon."

Parts List for 1984 911s

These are items most commonly replaced during regular maintenance and routine daily operation. Prices quoted are for new factory parts at list price, not including installation labor. NLA means factory parts no longer available, so prices quoted are from aftermarket suppliers.

Engine:

1. Oil filter..................... $18.52
2. Alternator belt........... $12.72
3. Starter....................... $433.15
4. Alternator $1,546.62
5. Muffler $1,510.93
6. Clutch disc............... $440.19

Body:

7. Front bumper............. $1,190.93
8. Left front fender........ $1,564.86
9. Right rear quarter panel............. $1,977.49
10. Front deck lid $1,865.68
11. Front deck lid struts.. $43.22 each
12. Rear deck lid struts... $32.88 each
13. Porsche badge, front deck lid............. $149.76
14. Taillight housing and lens $822.08
15. Windshield................. $1,011.43
16. Windshield weather stripping.................... $155.44

Interior:

17. Dashboard................. NLA $1,534.00
18. Shift knob $97.42
19. Interior carpet, complete (NLA) .... $400–$800

Chassis:

20. Front rotor................. $284.94
21. Brake pads, front set $95.04
22. Koni rear shock absorber.......... $213.50
23. Front wheel $1,374.67
24. Rear wheel $1,295.97

Ratings

1984 models, manual transmission

	911 Carrera coupe	911 Carrera Targa	911 Carrera Cabrio
Acceleration	4.5	4	3.5
Comfort	4.5	4.5	4
Handling	4.5	4	3.5
Parts	4.5	4b	4
Reliability	4q	4q	4q

b - Targa roofs no longer available from Porsche.
q - Alternator, see text.

1984 models, manual transmission, continued

	911 Carrera Turbo-Look coupe	911 Turbo
Acceleration	4.5	5p
Comfort	4.5	5p
Handling	4.5	3.5mp
Parts	4.5	1p
Reliability	4q	2pq

m - Turbo oversteer, see text
p - Turbo not sold in U.S. Gray-market imports only. See text.
q - Alternator, see text.

1984 Garage Watch
Problems with (and improvements to) Porsche 911 models.

Carrera! New 3.2-liter engine, new Digital Motor Electronics (DME) replaces CIS fuel injection, new oil-fed chain tensioner lube system virtually eliminates chain tensioner problems.

Heater blowers fail as from 1975; front fresh air blowers as well, as from 1969.

Targa roofs leak wind noise and water. Seals are very costly and hard to find, as from 1967.

Sunroof leaks, as from 1970.

Convertible tops, as from 1982.

Gray-market horrors, no U.S.-legal Turbo, but the Turbo look was optional.

New 1,050-watt alternator incorporates voltage regulator; however, has average life of only 40,000 miles.

Check engine compartment fuel line. Replacement is difficult job to do properly with engine in car; should remove engine. This problem continues through mid-year 1989.

Age problem appearing. The flexible portions of the fuel lines from the tank into the body—down the tunnel at the center of the car—then out of the tunnel to the engine, are beginning to rot from age and ozone exposure, causing fuel leaks at high pressure, especially in CIS cars, at about 75 psi.

Worn valve guides cause excess oil consumption.

Asbestos-free brake pads squeal; brake caliper piston O-rings deteriorate, become sticky.

148

Windshield and rear windows leak in corners when rubber seals get old and brittle, as from 1965.

Last year fender-mounted radio antenna.

Rust behind permanent side reflectors and in headlight buckets.

Brake master cylinders leak (see Chapter 23 for details).

Optional wider wheels/tires.

Electronic brake pad wear sensors introduced.

First and second gear synchros as from 1972.

Electric door locks introduced.

The Carrera 3.2 engine was introduced in model year 1984 (see Chapter 24 for details on the new engine).

U.S. emissions and safety standards continued to deny U.S. customers the Turbo (though it remained available in Canada). The gray-market importers and "federalizers" continued to bring in cars for U.S. customers under a one-time exemption. These operators ranged in quality from honest and responsible to shady with little engineering knowledge.

It is wise to avoid non-U.S. models. To identify them, look at the 17-digit VIN number. U.S. models read WPOAA with 12 numbers and letters after the double A. If you see WPOZZZ or WPOJ (Porsche's VIN prefix for its Canadian export cars), this is a non-U.S. model. Please see Chapter 3 and Chapters 21 through 24 for more information.

The gray market flourished from early calendar year 1980 through the late 1980s, even though Porsche began importing its Turbo model again in model year 1986.

Customers in Canada and the rest of the world could order a Turbo option from Rolf Sprenger's *Sonderwunsch*, or Special Wishes, department. Eventually designated the M505 for U.S. and M506 for rest-of-the-world customers, it was the *flachtbau* or flat nose steel body modification that first appeared in model year 1981. This took the flattened front fenders of Porsche's racing Typ 935 and mounted them on road cars. This would remain a special order option until model year 1987, a year after the Turbo would return to U.S. customers.

Understanding that as many owners wanted the look of the Turbo body as those seeking the additional performance, Porsche introduced option M491, the Turbo Look for Carrera coupes in model year 1984. For 1985, it included Targa and Cabrio models. This option added the Turbo's front and rear flared fenders to the standard Carrera body, as well as its front and rear spoilers. Less visible but more effective, the package fitted Turbo front hubs, rear suspension arms and torsion bar tube, the entire Turbo brake system, and the Turbo's wheels and tires.

The Carrera, with very few changes for model year 1985, continued to sell well. Porsche provided one-button central locking (already standard on the non-U.S. legal Turbos), and it replaced the fender-mounted radio antenna with one imbedded into the windshield. Porsche produced 11,859 Carreras in model year 1985. Total production exceeded 50,000 cars, and the U.S. took fully half, most of them 944 models as well as about 2,000 928 models.

1985 Specifications "F" Series

Body Designation:		911 Carrera
Price:		911 Carrera coupe: $31,960
		911 Carrera Targa: $31,450
		911 Carrera Cabriolet: $34,450
		Option M491 Turbo Look: $11,490
Engine Displacement and Type:		911SC (50 states)
		Typ 930/21; 3,164 cc (193.0 cid) SOHC, Bosch Digital Motor Electronic (DME)
		Motronic 2 Engine management system, Bosch LE-Jetronic fuel injection system
Maximum Horsepower @ rpm:		911 Carrera: 50 states:
		207 DIN @ 5,900 rpm 202 SAE net @ 5,900
Maximum Torque @ rpm:		911 Carrera: 50 states
		192 ft-lb DIN @ 4,000 rpm
		185 ft-lb SAE net @ 4,000 rpm
Weight:		911 Carrera: 2,670 pounds
		Cabrio: 2,750 pounds add 110 pounds for Turbo Look
0–60 mph:		911 Carrera: 6.7 seconds
		Cabrio: 7.0 seconds
Maximum Speed:		911 Carrera: 139 mph
		Cabrio: 134 mph, subtract 12 mph for Turbo Look (wind resistance)
Brakes:		vented disc brakes
Steering:		ZF rack-and-pinion
Suspension:	**Front:**	MacPherson struts with telescoping shock absorbers, wishbones, torsion bars, antisway bar
	Rear:	Telescoping shock absorbers, two-piece trailing arms, transverse torsion bars, antisway bar
Tires:	**Front:**	911 Carrera: 185/70HR15
		optional: 205/55VR16
	Rear:	911 Carrera: 215/60VR15
		optional: 225/50VR16
Tire air pressure:	**Front:**	28 psi; Rear: 35 psi
Transmission(s):		911 Carrera: 915/70 5-speed U.S.
Wheels:	**Front:**	911 Carrera: 6.0Jx15 Fuchs alloy
		optional: 7.0Jx16 Fuchs alloy
	Rear:	911 Carrera: 7.0Jx16 Fuchs alloy
		optional: 8.0Jx16 Fuchs alloy

What they said at the time–Porsche for 1985

911 Carrera, *Autosport*, April 1985

"Yes, the 911 is a magnificent car; yes it is still the most practical supercar around; yes it has an image second to none; yes it is one of the most exciting cars to drive made today. But it also has those niggling little flaws—the poor low-speed throttle response, odd minor control layout, nasty gearchange, awkward clutch and old-fashioned heating and ventilation—that are out of place in a 1985 car."

Parts List for 1985 911s

These are items most commonly replaced during regular maintenance and routine daily operation. Prices quoted are for new factory parts at list price, not including installation labor. NLA means factory parts no longer available, so prices quoted are from aftermarket suppliers.

Engine:

1. Oil filter.................... $18.52
2. Alternator belt.......... $16.00
3. Starter...................... $433.15
4. Alternator................. $1,073.55
5. Muffler..................... $1,510.93
6. Clutch disc............... $216.99

Body:

7. Front bumper............. $2,617.79
8. Left front fender........ $1,564.86
9. Right rear quarter panel............. $1,977.49
10. Front deck lid........... $1,865.68
11. Front deck lid struts.. $43.22 each
12. Rear deck lid struts... $32.88 each
13. Porsche badge, front deck lid............ $149.76
14. Taillight housing and lens................... $822.08
15. Windshield with antenna............. $1,011.43
16. Windshield weather stripping................... $115.44

Interior:

17. Dashboard (NLA)....... $1,534.00
18. Shift knob $97.42
19. Interior carpet, complete (NLA)........................ $400–$900

Chassis:

20. Front rotor................. $284.94
21. Brake pads, front set.................... $86.09
22. Koni rear shock absorber.......... $213.50
23. Front wheel.............. $1,374.67
24. Rear wheel............... $1,295.97

Ratings

1985 models, manual transmission

	911 Carrera coupe	911 Carrera Targa	911 Carrera Cabrio
Acceleration	4.5	4	4
Comfort	4.5	4	3.5
Handling	4.5	4	3.5
Parts	4.5	4b	4
Reliability	4q	4q	4q

b - Targa roofs no longer available from Porsche.
q - Alternator, see text.

1985 models, manual transmission, continued (2)

	Turbo-Look coupe	Turbo-Look Targa	Turbo Look Cabrio
Acceleration	4.5	4	4
Comfort	4.5	4	4
Handling	5	4.5	4
Parts	4.5	4b	4
Reliability	4q	4q	4q

b - Targa roofs no longer available from Porsche.
q - Alternator, see text.

1985 models, manual transmission, continued (3)

	911 Turbo coupe
Acceleration	5p
Comfort	5p
Handling	3.5mp
Parts	1ph
Reliability	2pq

m - Turbo oversteer, see text.   p - Turbo not sold in U.S. Gray-market imports only. See text.   q - Alternator, see text.

1985 Garage Watch
Problems with (and improvements to) Porsche 911 models.

Check engine compartment fuel line. Replacement is difficult job to do properly with engine in car; should remove engine. This problem continues through midyear 1989.

Convertible top mechanisms, as from 1982.

New 1,050-watt alternator incorporates voltage regulator; however, has average life of only 40,000 miles.

Valve Guides, as from 1984.

Brake caliper O-rings, as from 1984.

Buyers complain short shift kit option difficult to shift. Change of leverage point puts more work on driver's arm.

Gray-market horrors as from 1978; No U.S-legal Turbo, as from 1980.

This M491-option Turbo-Look coupe also has optional polished wheels.

Sunroof leaks, as from 1970.

Windshield radio antenna replaces fender mounted antenna.

Targa roofs leak wind noise and water. Seals are very costly and hard to find, as from 1967.

Windshield and rear windows leak in corners when rubber seals get old and brittle, as from 1965.

Heater blowers fail, as from 1975; front fresh air blowers continue to fail, as from 1969.

Rust behind permanent side reflectors, and in headlight buckets.

Electric seats introduced. Check motors for full operation in all positions.

Age problem appearing. The flexible portions of the fuel lines from the tank into the body—down the tunnel at the center of the car—then out of the tunnel to the engine, are beginning to rot from age and ozone exposure, causing fuel leaks at high pressure, especially in CIS cars, at about 75 psi.

Brake master cylinders leak. A problem common to all cars before 1990. The brakes would not fully release after being applied. This resulted from pedal bushings stiffening up or being swollen due to rust. The pistons stuck in the calipers. Brake hoses swelled shut, acting like a one-way valve. This problem disappeared with the introduction of power brakes.

1986

Chapter 26
The New
"G" Program—
The Turbo Is Back!

It took Bosch's Digital Motor Electronics (DME), oxygen sensors, and catalytic converters to make it work, but for model year 1986, Porsche reintroduced the 3.3-liter Turbo to U.S. markets, after an absence that started in model year 1980. At $48,000, they sold 1,424 of them.

An option first appeared in model year 1981 from Rolf Sprenger's *Sonderwunsch*, or Special Wishes, department, that was now available to U.S. customers. Eventually designated the M505 for U.S. and M506 for rest-of-the-world customers, it was the *flachtbau* or flat nose body modification for Turbocharged models. This took the flattened front fenders of Porsche's racing Typ 935 and mounted them on road cars. This would remain a special order option until model year 1987.

The Carrera 3.2 engine introduced in model year 1984 (see Chapter 24 for details).

The gray-market importers and "federalizers" continued to bring in cars for U.S. customers under a one-time exemption, even though Porsche reintroduced the Turbo for U.S. customers. These operators ranged in quality from honest and responsible to shady with little engineering knowledge.

Delays in getting Turbo models in the United States, or the desire to bring in a preowned model at a lower cost, kept these converters in business. It is wise to avoid non-U.S. models. To identify them, look at the 17-digit VIN number. U.S. models read WPOAA with 12 numbers and letters after the double A. If you see WPOJ (Porsche's prefix for Canadian models), or its rest-of-the-world prefix WPOZZZ, this is a non-U.S. model. You cannot get financing to purchase this car and you may not be able to insure it. The gray market flourished in the United States from early calendar year 1980 through the late 1980s.

Porsche upgraded the earlier braided fuel lines to seamless stainless steel, contoured to the engine and engine compartment. Limited stretches of braided line served as connections between the fixed stainless lines and the engine to allow for vibration.

Understanding that as many owners wanted the look of the Turbo body as those seeking the additional performance, Porsche introduced option M491, the Turbo Look for Carrera coupes in model year 1984. For 1985, it included Targa and Cabrio models. This option added the Turbo's front and rear flared fenders to the standard Carrera body, as well as its front and rear spoilers. Less visible but more effective, the package fitted Turbo front hubs, rear suspension arms and torsion bar tube, the entire Turbo brake system, and the Turbo's wheels and tires.

The Carrera, with very few changes for model year 1986, continued to sell well. Porsche improved its already popular Cabriolet model by introducing an optional power-lift mechanism. This was a complex system that used multiple motors to raise the top up and over top dead center, where momentum carried it to the other set of motors, which drew it down tightly. Because Porsche used cables to move the top bows, factory instructions strongly recommended raising or lowering the top only on a level surface to reduce the risk of misaligning the top and jamming it.

Porsche produced 12,696 Carreras in model year 1986. Total production remained above 50,000 cars, and the U.S. again took half, still mostly 944 models, as well as some 928 models.

1986 Specifications "G" Series

Body Designation:		911 Carrera, Turbo
Price:		911 Carrera coupe: $31,950
		911 Carrera Targa: $33,450
		911 Carrera Cabriolet: $36,450
		Option M491 Turbo Look: $11,490
		Turbo: $48,000
Engine Displacement and Type:		911 Carrera: Typ 930/21; 3,164 cc (192.6 cid) SOHC, Bosch Digital Motor Electronic (DME)
		Motronic 2 Engine management system, Bosch LE-Jetronic fuel injection system
		Turbo: Typ 930/68; 3,299 cc (201.3 cid) SOHC Bosch Digital Motor Electronic (DME)
		Motronic 2 Engine management system, Bosch LE-Jetronic fuel injection system, oxygen sensor and
		3-way Catalytic converters
Maximum Horsepower @ rpm:		911 Carrera: 50 states:
		202 SAE net @ 5,900 rpm
		Turbo: 282 SAE @ 5,500 rpm
Maximum Torque @ rpm:		911 Carrera: 50 states
		192 ft-lb @ 4,000 rpm
		Turbo: 287 @ 4,000
Weight:		911 Carrera: 2,670 pounds
		Cabrio: 2,750 pounds
		add 110 pounds for Turbo Look
		Turbo: 3,040 pounds
0–60 mph:		911 Carrera: 6.7 seconds
		Cabrio: 7.0 seconds
		Turbo: 4.6 (*Car and Driver*)
Maximum Speed:		911 Carrera: 139 mph
		Cabrio: 124 mph, subtract 12 mph for Turbo Look (wind resistance)
		Turbo: 155 mph (*Car and Driver*)
Brakes:		Carrera and Turbo: vented disc brakes
Steering:		ZF rack-and-pinion
Suspension:	Front:	MacPherson struts with telescoping shock absorbers, lower wishbones,
		longitudinal torsion bars, antisway bar
	Rear:	telescoping shock absorbers, two-piece trailing arms, transverse
		torsion bars, antisway bar
Tires:	Front:	911 Carrera: 185/70HR15
		optional: 205/55VR16
		Turbo: 225/50VR16
	Rear:	911 Carrera: 215/60VR15
		Optional: 225/50VR16
		Turbo: 245/45VR16
Tire air pressure:		Front 29 psi; Rear: 36 psi
		Turbo: Front: 29 psi; Rear: 43 psi
Transmission(s):		911 Carrera: 915/73 5-speed U.S.
		Turbo: 930/36 4-speed
Wheels:	Front	911 Carrera: 6.0Jx15 Fuchs alloy
		optional: 7.0Jx16 Fuchs alloy
		Turbo: 8.0Jx16 Fuchs alloy
	Rear:	911 Carrera: 7.0Jx16 Fuchs alloy
		optional: 8.0Jx16 Fuchs alloy
		Turbo: 9.0Jx16 Fuchs alloy

What they said at the time—Porsche for 1986

911 Turbo Slant Nose, *Car*, January 1986

"Beyond 4,000 rpm, if you are in a lower gear all hell breaks loose. . . Second is a remarkable gear. That one ratio encompasses the entire performance span of many lesser cars. It is possible to get the Porsche rolling in second. You can still be in second nearly 90 miles per hour later. Into the red, the speedo shows 95 miles per hour, but about 4–5 miles per hour of that you have to allow as speedo error. The car's sheer, thunderous performance has to be experienced to be believed."

911 Turbo, *Car and Driver*, January 1986

"To look at the new 911 Turbo is to stare right back into 1979. Only the keenest eye will notice that the rear tires now fill out the massive flared fenders a little more fully. Inside, the Turbo could be any 911 of recent vintage, but for a few minor details. A small boost gauge is incorporated into the tachometer at the six-o'clock position. Aside from that, the Turbo is just a well-dressed 911. Soft, sweet-smelling leather is lavished on the cockpit, including the dash top. A load of extras, from air conditioning to sunroof, are standard, just as you'd expect in a car that comes in at a nice, round $48,000."

Parts List for 1986 911s

These are items most commonly replaced during regular maintenance and routine daily operation. Prices quoted are for new factory parts at list price, not including installation labor. NLA means factory parts no longer available, so prices quoted are from aftermarket suppliers.

Engine:

1. Oil filter.................... $18.52
2. Alternator belt........... $16.00
3. Starter....................... $433.15
4. Alternator $1,073.55
5. Muffler $1,510.93
   (Turbo) $2,117.26
6. Clutch disc $440.16

Body:

7. Front bumper............. $2,617.79
8. Left front fender........ $1,584.86
   (Turbo fender
   plus flare) $1,891.93
9. Right rear
   quarter panel............. $1,730.50
10. Front deck lid $1,865.68
11. Front deck lid struts.. $43.22 each
12. Rear deck lid struts... $32.88 each
13. Porsche badge,
    front deck lid............. $149.76
14. Taillight housing
    and lens $822.08
15. Windshield
    with antenna $1,011.43
16. Windshield weather
    stripping.................... $155.44

Interior:

17. Dashboard (NLA)....... $1,534.00
18. Shift knob $97.42
19. Interior carpet, complete
    (NLA)......................... $400–$900

Chassis:

20. Front rotor................ $284.90
21. Brake pads,
    front set $86.09
22. Koni rear
    shock absorber $213.50
23. Front wheel $1,374.67
24. Rear wheel $1,295.97

Ratings

1986 models, manual transmission

	Carrera coupe	Carrera Targa	Carrera Cabrio
Acceleration	4	4	4
Comfort	4.5	4	4
Handling	4.5	4.5	4
Parts	4.5	4b	4
Reliability	4q	4q	4q

a - Targa roofs no longer available from Porsche.

q - Alternator, see text.

Ratings *continued*

1986 models, manual transmission, continued (2)

	Turbo coupe	Turbo Targa	Turbo Cabriolet
Acceleration	5	5	4.5
Comfort	5	4.5	4.5
Handling	4.5	4.5	4
Parts	4.5	4b	4
Reliability	4q	4q	4q

b - Targa roofs no longer available from Porsche.

q - Alternator, see text.

1986 models, manual transmission, continued

	Turbo-Look coupe	Turbo-Look Targa	Turbo-Look Cabrio
Acceleration	4	4	3.5
Comfort	4.5	4	4
Handling	5	4.5	4
Parts	4	3.5a	4
Reliability	4q	4q	4q

a - Targa roofs no longer available from Porsche.

q - Alternator, see text.

1986 Garage Watch
Problems with (and improvements to) Porsche 911 models.

Convertible tops get power option. Drive mechanism for the power top is vulnerable to slipping out of synch. Do not stop the stop mechanism part way through its operation. Don't take your finger off the switch. If it gets out of synch and refuses to close fully, do not force it. Attempting to force the top closed when this happens can severely damage the mechanisms and the top itself, sheering cables, and putting roof bows through top material.

Rear center stop light (Cyclops eye) at top of rear window.

First and second gear synchros, last year of 915 transmission, a problem as from 1972.

930 Turbo is back for U.S.

Check engine compartment fuel line and, after replacing it, reset CO. If the fuel line needs replacement it is a $500 part, plus installation. This problem continues through midyear 1989. This is difficult to do properly with engine in car; should remove engine.

Valve guides, as from 1984.

New 1,050-watt alternator incorporates voltage regulator; however, has average life of only 40,000 miles.

Short shift option, as from 1985.

Increased diameter anti-roll bar, rear brackets fail, making a resounding "clunk."

Age problem appearing. The flexible portions of the fuel lines from the tank into the body—down the tunnel at the center of the car—then out of the tunnel to the engine, are beginning to rot from age and ozone exposure, causing fuel leaks at high pressure, especially in CIS cars, at about 75 psi.

Brake caliper O-rings, as from 1984.

Sunroof leaks, as from 1970.

Targa roofs leak wind noise and water. Seals are very costly and hard to find, as from 1967.

Windshield and rear windows leak in corners when rubber seals get old and brittle, as from 1965.

Electric seats operation, as from 1985.

Heater blowers fail, as from 1975; front fresh air blowers continue to fail, as from 1969.

Brake master cylinders leak. A problem common to all cars before 1990. The brakes would not fully release after being applied. This resulted from pedal bushings stiffening up, or being swollen due to rust. The pistons stuck in the calipers. Brake hoses swelled shut, acting like a one-way valve. This problem disappeared with the introduction of power brakes.

For model year 1987, Weissach engineers remapped the Bosch Digital Motor Electronics engine management system on the Carrera 3.2 engine introduced in model year 1984. This boosted horsepower on the U.S. models to 217 horsepower SAE net (up from 202). U.S. buyers got a car that reached 60 miles per hour in 5.3 seconds and topped out at 152, performance just a hair slower than the 1976 Turbo.

The Bosch DME had replaced the continuous injection Bosch K-Jetronic fuel injection in model year 1984. This operated the fuel injection and ignition timing by monitoring throttle setting, outside and engine temperatures, engine load, engine speed, and even battery voltage. The DME "black box" sat under the driver's seat.

This new Carrera engine also addressed a problem dating from 1963. Porsche drove its overhead camshafts by chains. Oiling and keeping proper tensions on the chains had been a consistent problem. Ernst Fuhrmann, planning to end 911 production in 1983 or 1984, never authorized further engine improvements. Under Peter Schutz, Weissach engineers devised the Carrera chain tensioner and oiling system.

Porsche's Type 915 transmission also had held back further engine development. The gearbox was at the limits of its abilities to handle engine power and torque. Porsche switched to a Getrag five-speed, designated the G50, that used Borg-Warner synchronizers. Carreras and Turbos used the 240-millimeter hydraulic clutch. While this gearbox was capable of handling much more power, many have lamented its different feel as "a Camaro in a Porsche body."

Almost immediately, problems appeared with the hydraulic clutch. The factory issued a number of updates, to starter ring gear, throw-out fork, shaft and bearings, and guide tube and throw-out bearing clutches. These updates are complicated. However, it is unlikely that any cars still around have not had these improvements performed.

Porsche also had needed Bosch's Digital Motor Electronics (DME), oxygen sensors, and catalytic converters to make the 3.3-liter Turbo legal for U.S. markets beginning in model year 1986, after an absence beginning in model year 1980.

The Flat Nose Turbo body option that first appeared in model year 1981 from Rolf Sprenger's *Sonderwunsch*, or Special Wishes, department, was available worldwide. Designated the M506 for rest-of-the-world customers, Porsche fitted a large oil cooler in the nose. The U.S. Department of Transportation (USDOT) ruled this was a safety risk, and the M505 version for U.S. customers (beginning in March 1987) did not have the low-mounted oil cooler. Porsche designated this model the 930S. In all, Porsche produced 591 Slant Nose Turbos, in coupe, Cabrio, and Targa form in model year 1987. The company had resumed U.S. deliveries of the turbocharged cars beginning in model year 1986.

The gray-market importers and "federalizers" continued to bring in cars for U.S. customers under a one-time exemption, even though Porsche had already reintroduced the Turbo for U.S. customers. These operators ranged in quality from honest and responsible to shady with little engineering knowledge. There are many more questions than answers about the quality of individual conversions, and none of the converters are still in business to reply. Please see Chapter 3 for more information.

Delays in getting turbo models in the United States, or the desire to bring in a preowned model at a lower cost, kept these converters in business. It is wise to avoid non-U.S. models. To identify them, look at the 17-digit VIN number. U.S. models read WPOAA with 12 numbers and letters after the double A. If you see WPOJ, Porsche's prefix code for Canadian models, or WPOZZZ, the prefix it used for rest-of-the-world cars, this is a non-U.S. model. You cannot get financing to purchase this car and you may not be able to insure it. If you're after a car only for weekend outings with Porsche Owners' Club members on closed tracks, this would be an option for you. But for on-road use, you may find it's more trouble than it's worth. The gray market flourished in the United States from early calendar year 1980 through the late 1980s.

Porsche upgraded the earlier braided fuel lines to seamless stainless steel, contoured to the engine and engine compartment. Limited stretches of braided line served as connections between the fixed stainless lines and the engine to allow for vibration.

1987 Specifications "H" Series

Body Designation:		911 Carrera, Turbo
Price:		911 Carrera coupe: $31,950
		911 Carrera Targa: $33,450
		911 Carrera Cabriolet: $36,450
		Option: Slant Nose: $23,826
		Turbo: $48,000
Engine Displacement and Type:		Carrera: Typ 930/25; 3,164 cc (193.0 cid) SOHC, Bosch Digital Motor Electronic (DME) remapped Motronic 2 engine management system, Bosch LE-Jetronic fuel injection system.
		Turbo: Typ 930/68; 3,299 cc (201.3 cid) SOHC Bosch Digital Motor Electronic (DME) remapped Motronic 2 engine management system, Bosch LE-Jetronic fuel injection system, oxygen sensor and 3-way catalytic converters
Maximum Horsepower @ rpm:		911 Carrera: 50 states:
		217 SAE net @ 5,900 rpm
		Turbo: 282 SAE net @ 5,500 rpm
Maximum Torque @ rpm:		911 Carrera: 50 states 195 ft-lb @ 4,800 rpm
		Turbo: 288 ft-lb @ 4,000
Weight:		Carrera coupe: 2,756 pounds, add 110 pounds for Turbo Look
		Turbo: 2,976 pounds
0–60 mph:		911 Carrera: 6.1 seconds (factory)
		Cabrio: 7.0 seconds (factory)
		Turbo: 5.5 seconds (factory)
Maximum Speed:		911 Carrera: 149 mph (factory)
		Cabrio: 134 mph (factory), subtract 12 mph for Turbo Look (wind resistance)
		Turbo: 157 mph (factory)
Brakes:		Carrera and Turbo: vented disc brakes
Steering:		ZF rack-and-pinion
Suspension:	Front:	MacPherson struts with telescoping shock absorbers, wishbones, torsion bars, antisway bar
	Rear:	telescoping shock absorbers, two-piece trailing arms, transverse torsion bars, antisway bar
Tires:	Front:	911 Carrera: 185/70HR15
		optional: 205/55VR16
		Turbo: 205/55VR16
	Rear:	911 Carrera: 215/60VR15
		optional: 225/50VR16
		Turbo: 245/45VR16
Tire air pressure:	Front:	29 psi; Rear: 36 psi
		Turbo: Front: 29 psi; Rear: 43 psi
Transmission(s):		911 Carrera: 950/01 5-speed U.S.
		Turbo: 930/36 4-speed
Wheels:	Front:	911 Carrera: 6.0Jx15 Fuchs alloy
		optional: 7.0Jx16 Fuchs alloy
		Turbo: 7.0Jx16 Fuchs alloy
Rear:		911 Carrera: 7.0Jx16 Fuchs alloy
		optional: 8.0Jx16 Fuchs alloy
		Turbo: 9.0Jx16 Fuchs alloy

What they said at the time–Porsche for 1987

Turbo Slant Nose, *Car,* **January 1987**

"It's surprising how different the Turbo [Slant Nose] looks from an ordinary car. There's a less brutish, more exotic quality to it, and from the front more than a hint of 935 sports/racer. And that is much of what the buyer is paying for—a classier image for a car that goes as hard as any other production car on this earth up to 170 miles per hour."

Porsche 911 Cabriolet, *Car South Africa,* **January 1987**

"If the coupe's 'animal, almost primitive appeal' (to quote our own previous test) was a hard act to follow, the Cabriolet managed it with confident aplomb. Half our test team rate this car as the ultimate pleasure machine, no matter what factory came into the mix.

"For a car that's outrageously noisy, can by tricky in the wet, has dated if classic lines and borderline practicality, that may be surprising. The secret lies in the character of the car and its mighty power plant."

Parts List for 1987 911s

These are items most commonly replaced during regular maintenance and routine daily operation. Prices quoted are for new factory parts at list price, not including installation labor. NLA means factory parts no longer available, so prices quoted are from aftermarket suppliers.

Engine:

1. Oil filter.................... $18.52
2. Alternator belt.......... $16.00
3. Starter...................... $433.15
4. Alternator................. $1,073.55
5. Muffler..................... $1,510.93
   (Turbo) $1,000.17
6. Clutch disc............... $890.29

Body:

7. Front bumper............. $2,617.79
8. Left front fender........ $1,564.86
   (Turbo fender
   plus flare) $2,071.90
9. Right rear
   quarter panel............. $1,730.50
10. Front deck lid........... $1,865.68
11. Front deck lid struts.. $43.22 each
12. Rear deck lid struts... $32.88 each
13. Porsche badge,
    front deck lid............. $149.76
14. Taillight housing
    and lens.................... $822.08
15. Windshield
    with antenna $1,011.43
16. Windshield weather
    stripping.................... $115.44

Interior:

17. Dashboard (NLA)....... $1,534.00
18. Shift knob $97.42
19. Interior carpet, complete
    (NLA)........................ $400–$900

Chassis:

20. Front rotor................. $284.94
21. Brake pads,
    front set $86.09
22. Koni rear
    shock absorber $213.50
23. Front wheel $1,374.67
24. Rear wheel $1,295.97

Ratings

1987 models, manual transmission

	Carrera coupe	Carrera Targa	Carrera Cabrio
Acceleration	4r	4r	3.5r
Comfort	4.5	4.5	4.5t
Handling	4.5	4.5	4
Parts	4.5	4b	4.5
Reliability	3.5qrs	3.5qrs	3qrst

b - Targa roofs no longer available from Porsche.
q - Alternator, see text
r - New transmission, clutch, see text.
s - Fuel lines, see text.
t - Power top, improves comfort, risks jamming, see text.

Ratings *continued*

1987 models, manual transmission, continued (2)

	Turbo coupe	Turbo Targa	Turbo Cabriolet
Acceleration	5r	5r	4.5r
Comfort	5	4.5	5t
Handling	4	4	3.5
Parts	4.5	4b	4.5
Reliability	3.5qrs	3.5qrs	3qrst

b - Targa roofs no longer available from Porsche.
q - Alternator, see text.
r - New transmission, clutch, see text.
s - Fuel lines, see text.
t - Power top, improves comfort, risks jamming, see text.

1987 models, manual transmission, continued (3)

	Slant-Nose Turbo coupe	Slant Nose Targa	Slant Nose Cabrio
Acceleration	5r	5r	4.5r
Comfort	5	4.5	5t
Handling	4.5	4	4
Parts	3.5u	3au	3.5u
Reliability	3.5qrs	3.5qrs	3qrst

b - Targa roofs no longer available from Porsche.
q - Alternator, see text
r - New transmission, clutch, see text.
s - Fuel lines, see text.
t - Power top, improves comfort, risks jamming, see text.
u - Factory Slant Nose steel body panels extremely costly.

1987 models, manual transmission, continued (4)

	Turbo-Look coupe	Turbo-Look Targa	Turbo-Look Cabrio
Acceleration	4r	4r	3.5r
Comfort	4.5	4	4.5t
Handling	5	4.5	4
Parts	4	3.5b	4
Reliability	3.5qrs	3.5qrs	3qrst

b - Targa roofs no longer available from Porsche.
q - Alternator, see text
r - New transmission, clutch, see text.
s - Fuel lines, see text.
t - Power top, improves comfort, risks jamming, see text.

These flexible portions of the fuel lines, from the tank into the body (they are stainless tubing through the center tunnel) and out of the tunnel to the engine, are beginning to rot from age and ozone exposure.

Cars built in model years 1987 and 1988 are more prone to suffer this problem. Porsche changed outside manufacturers for these lines, and this particular firm failed to meet Porsche's quality requirements.

This causes fuel leaks at high pressure. (The CIS Bosch K-Jetronic pumps fuel at 75 psi.) These leaks can only be seen from underneath the car, an excellent reason to have a prepurchase inspection. If you smell fuel or see a puddle, do not start the car; there is a fire risk.

In order to meet the growing electrical requirements of electric windows, roofs, air conditioning, and security systems (and electric seats in model year 1984), Porsche fitted a new 1,050-watt alternator that incorporated a voltage regulator on SC and Carrera models through model year 1989. Unfortunately, this new unit proved to have only a 40,000-mile life. As it approaches failure, it overcharges the battery. Overcharging makes the headlights surge,

the air conditioner blower fan speed up and slow down, and the battery stink. If a noxious odor begins to pervade the passenger compartment from the front trunk compartment, this likely means the alternator is overcharging the battery and it is beginning to fail. The alternator will spike voltage, burning out other electric components and incurring costly repairs and replacements.

Understanding that as many owners wanted the look of the Turbo body as those seeking the additional performance, Porsche introduced option M491, the Turbo Look for Carrera coupes in model year 1984. For 1985, they included Targa and Cabrio models. This option added the Turbo's front and rear flared fenders to the standard Carrera body, as well as its front and rear spoilers. Less visible but more effective, the package fitted Turbo front hubs, rear suspension arms and torsion bar tube, the entire Turbo brake system, and the Turbo's wheels and tires.

The Carrera, with its substantial changes for model year 1987, including a standard power top for the Cabrio, continued to sell well. Porsche produced 14,853 Carreras, and 1,727 Turbos in all forms in model year 1987.

1987 Garage Watch
Problems with (and improvements to) Porsche 911 models.

Convertible tops get power option with drive mechanism vulnerable to slipping out of synch, as from 1986.

Sunroof leaks, as from 1970.

Third brake light "Cyclops eye" relocated to bottom of rear window. This owner removed "Cyclops" and replaced it with later-series interior unit.

New G50 five-speed transmission ends first and second gear synchro failures. If you shift too quickly into third gear, you crunch. For this transmission, many service updates need to be performed. These are complicated and include the starter ring gear, throw-out fork and bearings, and the guide tube and throw-out bearing clutches as they age.

Check engine compartment fuel line and, after replacing it, reset CO. If the fuel line needs replacement it is a $500 part, plus installation. This problem continues through midyear 1989 but was worst in 1987 and 1988, due to outside supplier who manufactured lines for Porsche. This is difficult to do properly with engine in car; should remove engine, as from 1984.

Valve guides fail, as from 1984, but are more prone to fail because Porsche used particular phosphorous bronze materials for part of 1987 and 1988 model years. It is likely, however, that these will have been replaced. This is something to look for in owner's work receipts.

New 1,050-watt alternator incorporates voltage regulator; however, has average life of only 40,000 miles.

As hydraulic clutch gets older, it gets stiffer to operate. With 1987 and 1988 models especially, if the clutch is updated, driver ends up with buttery-soft clutch that is wonderful to drive.

Brake caliper O-rings leak, as from 1984.

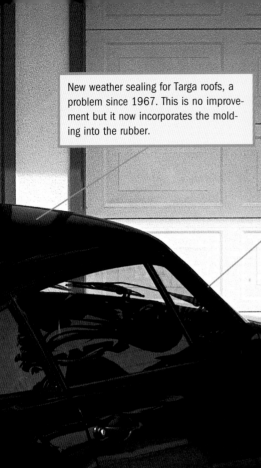

New weather sealing for Targa roofs, a problem since 1967. This is no improvement but it now incorporates the molding into the rubber.

Windshield and rear windows leak in corners when rubber seals get old and brittle, as from 1965.

Heater blowers fail, as from 1975; front fresh air blowers continue to fail, as from 1969.

Steel Slant Nose option for Turbos.

Halogen headlights brighter but now require bulbs.

Electric seats operation, as from 1985.

Brake master cylinders leak. A problem common to all cars before 1990. The brakes would not fully release after being applied. This resulted from pedal bushings stiffening up, or being swollen due to rust. On cars built in 1976 and earlier, the master cylinder seeped brake fluid onto the bushings, and it leaked out into the pedal assembly. The pistons stuck in the calipers. Brake hoses swelled shut, acting like a one-way valve. This problem disappeared with the introduction of power brakes and its resulting complete redesign of the brake system.

Age problem appearing. The flexible portions of the fuel lines from the tank into the body—down the tunnel at the center of the car—then out of the tunnel to the engine, are beginning to rot from age and ozone exposure, causing fuel leaks at high pressure, especially in CIS cars, at about 75 psi.

Larger rear anti-roll bar brackets break, listen for "clunk," as from 1986.

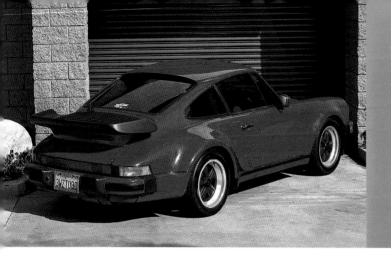

1988

Chapter 28
The New "J" Program—Small Changes While New Models Lurk

NOTE: The 1988 model was virtually unchanged from 1987.

For model year 1987, Weissach engineers remapped the Bosch Digital Motor Electronics engine management system on the Carrera 3.2 engine introduced in model year 1984. This boosted horsepower and torque on the U.S. models to 217 horsepower SAE net, though still at 231 DIN at 5,900 rpm and 195 ft-lb SAE net, 209 DIN at 4,800 rpm. U.S. buyers got a car that reached 60 miles per hour in 5.3 seconds and topped out at 152, performance just a hair slower than the 1976 Turbo (see Chapter 27 for details).

The Bosch DME had replaced the continuous injection Bosch K-Jetronic fuel injection. This operated the fuel injection and ignition timing by monitoring throttle setting, outside and engine temperatures, engine load, engine speed, and even battery voltage. Porsche mounted the DME "black box" under the driver's seat.

This new Carrera engine also addressed a problem dating from 1963. Porsche drove its overhead camshafts by chains. Oiling and keeping proper tensions on the chains had been a consistent problem. Ernst Fuhrmann, set on ending 911 production in 1983 or 1984, never authorized further improvements. Under Peter Schutz, Weissach engineers devised the new Carrera chain tensioner and oiling system.

Porsche's Type 915 transmission also had held back further engine development. The gearbox was at the limits of its abilities to handle engine power and torque. In model year 1987, Porsche introduced the G50, a Getrag five-speed that used Borg-Warner synchronizers. While this gearbox was capable of handling much more power, several sources consulted for this book have lamented its different feel, describing it as "a Camaro in a Porsche body."

Almost immediately, problems appeared with the hydraulic clutch. The factory issued a number of updates, to starter ring gear, throw-out fork, shaft and bearings, and guide tube and throw-out bearing clutches. These updates are complicated. However, it is unlikely that any cars still around have not had these improvements performed.

Porsche continued worldwide sales of its 3.3-liter Turbo, available with the Flat Nose Turbo body option. This M506 for rest-of-the-world customers featured a large oil cooler in the nose. The U.S. Department of Transportation (USDOT) ruled this was a safety risk, and the M505 version for U.S. customers (beginning in March 1987) did not have the low-mounted oil cooler. Porsche designated the Slant Nose Turbo model the 930S.

The gray-market importers and "federalizers" continued to bring in cars for U.S. customers under a one-time exemption, even though Porsche had already reintroduced the Turbo for U.S. customers. See Chapter 3 for more information.

Porsche upgraded the earlier braided fuel lines to seamless stainless steel, contoured to the engine and engine compartment. Limited stretches of braided line served as connections between the fixed stainless lines and the engine to allow for vibration.

These flexible portions of the fuel lines, from the tank into the body (they are stainless tubing through the center tunnel) and out of the tunnel to the engine, are beginning to rot from age and ozone exposure.

Cars built in model years 1987 and 1988 are more prone to suffer this problem. Porsche changed outside manufacturers for these lines and this firm failed to meet quality requirements.

This causes fuel leaks at high pressure. (The CIS Bosch K-Jetronic pumps fuel at 75 psi.) These leaks can only be seen from underneath the car, an excellent reason to have a prepurchase inspection. If you smell fuel or see a puddle, do not start the car; there is a fire risk.

Beginning in model year 1984, to meet the growing electrical requirements of electric windows, roofs, air conditioning and security systems (and electric seats in model year 1984), Porsche fitted a 1,050-watt alternator that incorporated a voltage regulator on SC and Carrera models through model year 1989. Unfortunately, this new unit proved to have only a 40,000-mile life. As it approaches failure, it overcharges the battery. Overcharging makes the headlights surge, the air conditioner blower fan speed up and slow down, and the battery stink. If a noxious odor begins to pervade the passenger compartment from the front trunk compartment, this likely means the alternator is overcharging the battery and it is beginning

1988 Specifications "J" Series

Body Designation:		911 Carrera, Turbo
Price:		911 Carrera coupe: $45,895
		911 Carrera Club Sport: $48,895
		Turbo: $68,670
		Slant Nose option: $23,826 (plus car)
Engine Displacement and Type:		911 Carrera: Typ 930/25; 3,164 cc (193.0 cid) SOHC Bosch DME Digital Motor Electronic remapped Motronic 2 engine management system, Bosch LE-Jetronic fuel injection system
		Turbo: Typ 930/68; 3,299 cc (201.3 cid) SOHC Bosch DME Digital Motor Electronic remapped Motronic 2 engine management system, Bosch LE-Jetronic fuel injection system, oxygen sensor and 3-way catalytic converters, KKK turbocharger
Maximum Horsepower @ rpm:		911 Carrera: 50 states: 217 SAE net @ 5,900 rpm
		Turbo: 282 SAE net @ 5,500 rpm
Maximum Torque @ rpm:		911 Carrera: 50 states 195 ft-lb @ 4,800 rpm
		Turbo: 287ft-lb @ 4,000
Weight:		911 Carrera: 2,756 pounds
		911 Club Sport: 2,601 pounds
		Turbo: 2,976 pounds, add 110 pounds for Turbo Look Carrera
0–60 mph:		911 Carrera: 6.1 seconds (factory)
		911 Club Sport: 6.2 seconds (*Car and Driver*)
		Cabrio: 7.0 seconds (factory)
		Turbo: 5.5 seconds (factory)
Maximum Speed:		911 Carrera: 149 mph (factory)
		911 Club Sport: 152 mph (*Car and Driver*)
		Cabrio: 134 mph (factory), subtract 12 mph for Turbo Look (wind resistance)
		Turbo: 157 mph (factory)
Brakes:		Carrera: power vented discs
		Turbo: power vented and cross-drilled discs
Steering:		ZF rack-and-pinion
Suspension:	Front:	MacPherson struts, torsion bars
	Rear:	semi-trailing arms, transverse torsion bars, antisway bar
Tires:	Front:	911 Carrera: 185/70HR15
		optional: 205/55VR16
		Turbo: 225/50VR16
	Rear:	911 Carrera: 215/60VR15
		optional: 225/50VR16
		Turbo: 245/45VR16
Tire air pressure:	Front:	29 psi; Rear: 34 psi
		Turbo: Front: 29 psi; Rear: 43 psi
Transmission(s):		911 Carrera: G50/01 5-speed U.S.
		Turbo: G50/50 5-speed
Wheels:	Front:	911 Carrera: 6.0Jx15 Fuchs alloy
		optional: 7.0Jx16 Fuchs alloy
		Turbo: 8.0Jx16 Fuchs alloy
	Rear:	911 Carrera: 7.0Jx16 Fuchs alloy
		optional: 8.0Jx16 Fuchs alloy
		Turbo: 9.0Jx16 Fuchs alloy

What they said at the time–Porsche for 1988

Porsche 911 Turbo [Slant Nose] Cabriolet, *Car and Driver,* September 1987

"The nose job arrived during the 1976 season: Porsche managed to slip a 935 with a new leading edge through a loophole in the Group 5 rules. Lopping off the traditional fenders and headlamps produced the now fabled Slant Nose shape, which contributed much more front downforce. It was a short step to modifying production 911 Turbos with bodywork that resembled the 935 racers, and soon a number of German customs houses and various American specialists were cranking out streetable replicas.

"By 1981, it was painfully clear that Stuttgart was forfeiting a rollicking good business to the aftermarket. Ever since [setting up Special Wishes] Porsche has been in the custom biz with a vengeance, handcrafting megadollar conversions for upscale customers."

Porsche 911 Club Sport, *Motor Trend,* August 1988

"Those of us with a cynical nature also have to wonder if a bit of clever marketing hasn't also found its way into the Club Sport formula. If you order the Club Sport, you don't have to hire a high-priced race mechanic to pull out the air-conditioner and ash can the rear seats. Removing the passenger visor alone could cost some poor sparrow of a race car owner $100 in labor. . . ."

Parts List for 1988 911s

These are items most commonly replaced during regular maintenance and routine daily operation. Prices quoted are for new factory parts at list price, not including installation labor. NLA means factory parts no longer available, so prices quoted are from aftermarket suppliers.

Engine:

1. Oil filter.................... $18.52
2. Alternator belt........... $16.00
3. Starter....................... $433.15
4. Alternator................. $1,073.55
5. Muffler...................... $1,510.93
   (Turbo) $2,117.26
6. Clutch disc............... $890.29

Body:

7. Front bumper............. $2,617.79
8. Left front fender........ $1,564.86
   (Turbo fender
   plus flare) $2,071.90
9. Right rear
   quarter panel............. $1,977.49
10. Front deck lid........... $1,865.68
11. Front deck lid struts.. $43.22 each
12. Rear deck lid struts... $32.88 each
13. Porsche badge,
    front deck lid............. $149.76
14. Taillight housing
    and lens................... $822.08
15. Windshield
    with antenna............. $1,011.43
16. Windshield weather
    stripping................... $115.44

Interior:

17. Dashboard (NLA)....... $1,534.00
18. Shift knob $97.42
19. Interior carpet,
    complete (NLA) .... $400–$900

Chassis:

20. Front rotor................. $284.94
21. Brake pads,
    front set.................... $86.09
22. Koni rear
    shock absorber.......... $213.50
23. Front wheel $1,374.67
24. Rear wheel $1,295.97

Ratings

1988 models, manual transmission

	Carrera coupe	Carrera Targa	Carrera Cabrio
Acceleration	4r	4r	3.5r
Comfort	4.5	4	4.5t
Handling	4.5	4.5	4
Parts	4	3.5b	4
Reliability	3.5qrs	3.5qrs	3qrst

b - Targa roofs no longer available from Porsche.
q - Alternator, see text.
r - New transmission, clutch, see text.
s - Fuel lines, see text.
t - Power top, improves comfort, risks jamming, see text.

1988 models, manual transmission, continued (2)

	Turbo coupe	Turbo Targa	Turbo Cabriolet
Acceleration	5r	4.5r	4.5r
Comfort	5	4.5	5t
Handling	4	4	4
Parts	4.5	4a	4.5
Reliability	3.5qrs	3.5qr	3qrst

b - Targa roofs no longer available from Porsche.
q - Alternator, see text.
r - New transmission, clutch, see text.
s - Fuel lines, see text.
t - Power top, improves comfort, risks jamming, see text.

Ratings

1988 models, manual transmission, continued (3)

	Slant Nose Turbo coupe	Slant Nose Targa	Slant Nose Cabrio
Acceleration	5r	5r	4.5r
Comfort	5	4.5	5t
Handling	4.5	4	4
Parts	3.5u	3.5bu	3.5u
Reliability	3.5qrs	3.5qrs	3qrst

b - Targa roofs no longer available from Porsche.
q - Alternator, see text.
r - New transmission, clutch, see text.
s - Fuel lines, see text.
t - Power top, improves comfort, risks jamming, see text.
u - Factory Slant Nose steel body panels extremely costly.

1988 models, manual transmission, continued (4)

	Turbo-Look coupe	Turbo-Look Targa	Turbo-Look Cabrio
Acceleration	4r	4r	3.5r
Comfort	4.5	4	4.5t
Handling	5	4.5	4
Parts	4	3.5b	4
Reliability	3.5qrs	3.5qrs	3qrst

b - Targa roofs no longer available from Porsche.
q - Alternator, see text.
r - New transmission, clutch, see text.
s - Fuel lines, see text.
t - Power top, improves comfort, risks jamming, see text.

to fail. The alternator will spike voltage, burning out other electric components and incurring costly repairs and replacements.

Lastly, Porsche produced a limited run of 911 Club Sport models. This was a severely stripped model, lacking sound insulation, air conditioner, door panels, and other accouterments, deletions meant to pull 155 pounds off the standard Carrera.

Bosch DME tuning runs the redline to 6,850 rpm instead of 6,520 but the engine still produced 214 SAE net horsepower at 5,900. Porsche fitted the flat whale tail wing and removed the front fog lights, seemingly as preparation for . . . umm . . . club racing requirements. It sold for $48,895, same as the fully equipped Carrera coupe.

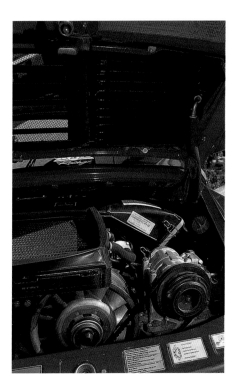

1988 Garage Watch
Problems with (and improvements to) Porsche 911 models.

Special model Club Sport introduced.

Convertible tops, as from 1986.

Windshield and rear windows leak in corners when rubber seals get old and brittle, as from 1965.

Heater blowers fail, as from 1975; front fresh air blowers continue to fail, as from 1969.

Electric seats operations, as from 1985.

As hydraulic clutch gets older, it gets stiffer to operate. With 1987 and 1988 models especially, if the clutch is updated, driver ends up with buttery-soft clutch that is wonderful to drive.

Brake caliper O-rings, as from 1984.

Sunroof leaks, as from 1970.

New weather sealing for Targa roofs, a problem since 1967. This is no improvement but it now incorporates the molding into the rubber.

Cyclops eye back up to top of rear window.

Check engine compartment fuel line and, after replacing it, reset CO. If the fuel line needs replacement it is a $500 part, plus installation. This problem continues through midyear 1989, but was worst in 1987 and 1988 due to outside supplier who manufactured lines for Porsche. This job is difficult to do properly with engine still in car; should remove engine, as from 1984.

Valve guides, as from 1984.

1,050-watt alternator incorporates voltage regulator; however, has average life of only 40,000 miles.

For this transmission, many service updates need to be performed. These are complicated and include the starter ring gear, throw-out fork and bearings, and the guide tube and throw-out bearing clutches as they age.

Brake master cylinders leak. A problem common to all cars before 1990. The brakes would not fully release after being applied. This resulted from pedal bushings stiffening up, or being swollen due to rust. The pistons stuck in the calipers. Brake hoses swelled shut, acting like a one-way valve. This problem disappeared with the introduction of power brakes.

Age problem appearing. The flexible portions of the fuel lines from the tank into the body—down the tunnel at the center of the car—then out of the tunnel to the engine, are beginning to rot from age and ozone exposure, causing fuel leaks at high pressure, especially in CIS cars, at about 75 psi.

NOTE: The 1989 model was virtually identical to the 1988 and 1987 models.

For model year 1987, Weissach engineers had remapped the Bosch Digital Motor Electronics engine management system on the Carrera 3.2 engine introduced in model year 1984, boosting horsepower and torque on U.S. models to 217 horsepower SAE net, though still at 231 DIN at 5,900 rpm and 195 ft-lb SAE net, 209 DIN at 4,800 rpm.

The Bosch DME had replaced the continuous injection Bosch K-Jetronic fuel injection. This operated the fuel injection and ignition timing by monitoring throttle setting, outside and engine temperatures, engine load, engine speed, and even battery voltage. The DME "black box" sat under the driver's seat.

This new Carrera engine also addressed a problem dating from 1963. Porsche drove its overhead camshafts by chains. Oiling and keeping proper tensions on the chains had been a consistent problem. Ernst Fuhrmann, looking to end 911 production in 1983 or 1984, never authorized further improvements. Under Peter Schutz, Weissach engineers devised the Carrera chain tensioner and oiling system.

Porsche's Type 915 transmission also had held back further engine development. The gearbox was at the limits of its abilities to handle engine power and torque. In model year 1987, Porsche introduced the G50, a Getrag five-speed that used Borg-Warner synchronizers. While this gearbox was capable of handling much more power, many have lamented its different feel as "a Camaro in a Porsche body."

Almost immediately, problems appeared with the hydraulic clutch. The factory issued a number of updates, to starter ring gear, throw-out fork, shaft and bearings, and guide tube and throw-out bearing clutches. These updates were complicated, but for model year 1989, Porsche performed the updates before delivery.

Porsche continued worldwide sales of its 3.3-liter Turbo, available with the Flat Nose Turbo body option. This M506 for rest-of-the-world customers featured a large oil cooler in the nose. The U.S. Department of Transportation (USDOT) ruled this was a safety risk, and the M505 version for U.S. customers (beginning in March 1987) did not have the low-mounted oil cooler. Porsche designated this Slant Nose Turbo model the 930S.

Porsche upgraded the earlier braided fuel lines to seamless stainless steel, contoured to the engine and engine compartment. Limited stretches of braided line served as connections between the fixed stainless lines and the engine to allow for vibration.

These flexible portions of the fuel lines, from the tank into the body (they are stainless tubing through the center tunnel) and out of the tunnel to the engine, are beginning to rot from age and ozone exposure. This causes fuel leaks at high pressure. These leaks can only be seen from underneath the car, an excellent reason to have a prepurchase inspection. If you smell fuel or see a puddle, do not start the car; there is a fire risk.

In order to meet the growing electrical requirements of electric windows, roofs, air conditioning, and security systems (and electric seats in model year 1984), Porsche fitted a 1,050-watt alternator that incorporated a voltage regulator on SC and Carrera models through model year 1989. Unfortunately, this new unit proved to have only a 40,000-mile life. As it approaches failure, it overcharges the battery. Overcharging makes the headlights surge, the air conditioner blower fan speed up and slow down, and the battery stink. If a noxious odor begins to pervade the passenger compartment from the front trunk compartment, this likely means the alternator is overcharging the battery and it is beginning to fail. The alternator will spike voltage, burning out other electric components and incurring costly repairs and replacements.

For model year 1989, Porsche introduced a car for which President Peter Schutz and chief engineer Helmuth Bott had special fondness, the Speedster. It was blatantly meant to remind long-time Porsche owners and enthusiasts of the late 1950s models meant to tantalize California buyers. Just as buyers' affections are divided over Porsche's Targas, and even its Cabriolets, the Speedster was an acquired taste. Some people hated its lines from any angle, others swooned, and then wrote checks. Styling chief Tony Lapine pulled visual cues from the 356 models and carried over the lower, raked-back windscreen, and added the most controversial element of all, the "twin camel humps" that covered the rear luggage area—the seats were removed entirely. Lapine and

1989 Specifications "K" Series

Body Designation: 911 Carrera, Speedster, Turbo

Price:
911 Carrera coupe: $51,205
911 Carrera Targa: $53,450
911 Carrera Cabriolet: $56,450
Speedster: $65,480
Turbo: $70,975

Engine Displacement and Type: 911 Carrera: Typ 930/25; 3,164 cc (193.0 cid) SOHC Bosch DME Digital Motor Electronic remapped Motronic 2 engine management system, Bosch LE-Jetronic fuel injection system, 3-way catalytic converter, oxygen sensor.
Turbo: Typ 930/68; 3,299 cc (201.3 cid) SOHC Bosch K-Jetronic fuel injection system, KKK turbocharger and air-to-air intercooler, 3-way catalytic converter with oxygen sensor and secondary air injection

Maximum Horsepower @ rpm: 911 Carrera: 50 states:
214 SAE net @ 5,900 rpm
Turbo: 282 SAE net @ 5,500 rpm

Maximum Torque @ rpm: 911 Carrera: 50 states
195 ft-lb @ 4,800 rpm
Turbo: 288 @ 4,000

Weight: Carrera coupe: 2,756 pounds
Cabrio: 2,916 pounds add 110 pounds for Turbo Look
Speedster: 2,756 pounds
Turbo: 2,976 pounds

0–60 mph: 911 Carrera: 6.1 seconds (factory)
Turbo: 5.3 seconds (factory)

Maximum Speed: 911 Carrera: 149 mph (factory)
Cabrio: 134 mph (top down) (*Car and Driver*)
subtract 12 mph for Turbo Look (wind resistance)
Speedster: 137 mph (factory)
Turbo: 157 mph (factory)

Brakes: Carrera: power vented discs
Turbo: power vented and cross-drilled discs

Steering: ZF rack-and-pinion

Suspension: Front: MacPherson struts, torsion bars, antisway bar
Rear: semi-trailing arms, transverse torsion bar, antisway bar

Tires: Front: 911 Carrera: 205/55VR16
Turbo: 205/55VR16
Rear: 911 Carrera: 225/50VR16
Turbo: 245/45VR16

Tire air pressure: Front: 29 psi; Rear: 36 psi
Turbo: Front: 29 psi; Rear: 43 psi

Transmission(s): 911 Carrera: G50/01 5-speed U.S.
Turbo: G50/50 5-speed

Wheels: Front: 911 Carrera: 6.0Jx16 Fuchs alloy
Turbo: 7.0Jx16 Fuchs alloy
Rear: 911 Carrera: 8.0Jx16 Fuchs alloy
Turbo: 9.0Jx16 Fuchs alloy

What they said at the time—Porsche for 1989

1989 Porsche Carrera, *Road & Track*, May 1989

"Some [recent] family sedans will generate performance numbers that a 911 Carrera can't match. Yet people buy these quarter-century old favorites. Time has not diminished the Carrera's appeal, or changed it much.

"That's good, because the 1989 Carrera is in something of a holding pattern. For the third year in a row, no modifications have been made."

1989 Porsche Turbo, *Road & Track*, May 1989

"Though the new [five-speed] gearbox hardly makes a difference to standing start acceleration times, it makes the car much more pleasant and relaxing to drive, as there is no need to rev the engine near redline to keep it 'on song' when the next higher gear is selected. With third good for 109 miles per hour and fourth reaching over 130 before fifth is needed, the gears are now nicely stepped."

1989 Carrera Speedster, *Motor Trend*, January 1990

"We have in the Speedster a bit of an anomaly: It's built on the old 911 chassis, yet even at list price, it costs about the same as the much more advanced Carrera 2 Cabriolet that will be available this winter. Consider the Speedster's drawbacks: It's got less displacement, less power, no ABS, no power steering, no back seat, it leaks in the rain and it has an awkward manual top. Yet the people who pay well over list price for the car know exactly what they're buying into."

Parts List for 1989 911s Carrera

These are items most commonly replaced during regular maintenance and routine daily operation. Prices quoted are for new factory parts at list price, not including installation labor. NLA means factory parts no longer available, so prices quoted are from aftermarket suppliers.

Engine:

1. Oil filter $15.82
2. Alternator belt $16.00
3. Starter $433.15
4. Alternator $1,073.55
5. Muffler $1,510.93
   (Turbo) $2,117.26
6. Clutch disc $890.29

Body:

7. Front bumper $2,617.79
8. Left front fender $1,564.86
   (Turbo fender
   plus flare) $2,071.90
9. Right rear
   quarter panel $1,730.50
10. Front deck lid $1,865.68
11. Front deck lid struts .. $43.22 each
12. Rear deck lid struts ... $32.88 each
13. Porsche badge,
    front deck lid $149.76
14. Taillight housing
    and lens $822.08
15. Windshield
    with antenna $1,011.43
16. Windshield weather
    stripping $155.44

Interior:

17. Dashboard (NLA) $1,534.00
18. Shift knob $97.42
19. Interior carpet,
    complete (NLA) .... $400–$900

Chassis:

20. Front rotor $284.90
21. Brake pads,
    front set $86.09
22. Koni rear
    shock absorber $213.50
23. Front wheel $1,374.67
24. Rear wheel $1,295.97

Ratings

1989 models, manual transmission

	911 Carrera coupe	911 Carrera Targa	911 Carrera Cabrio
Acceleration	4r	4r	3.5r
Comfort	4.5	4	4.5t
Handling	4.5	4.5	4
Parts	4	3.5b	4
Reliability	3.5qrs	3.5qrs	3qrst

b - Targa roofs no longer available from Porsche.
q - Alternator, see text.
r - New transmission, clutch, see text.
s - Fuel lines, see text.
t - Power top, improves comfort, risks jamming, see text.

Ratings <small>continued</small>

1989 models, manual transmission, continued (2)

	Turbo coupe	Turbo Targa	Turbo Cabriolet
Acceleration	5r	5r	4.5r
Comfort	5	4.5	4.5t
Handling	4.5	4	4
Parts	4	3.5b	4
Reliability	3.5qrs	3.5qrs	3qrst

b- Targa roofs no longer available from Porsche.
q - Alternator, see text.
r - New transmission, clutch, see text.
s - Fuel lines, see text.
t - Power top, improves comfort, risks jamming, see text.

1989 models, manual transmission, continued (3)

	Slant-Nose Turbo coupe	Slant Nose Targa	Slant Nose Cabrio
Acceleration	5r	5r	4.5r
Comfort	5	4.5	5t
Handling	4.5	4	4
Parts	3.5u	3.5bu	3.5u
Reliability	3.5qrs	3.5qrs	3qrst

b - Targa roofs no longer available from Porsche.
q - Alternator, see text.
r - New transmission, clutch, see text.
s - Fuel lines, see text.
t - Power top, improves comfort, risks jamming, see text.
u - Factory Slant Nose steel body panels extremely costly.

1989 models, manual transmission, continued (4)

	Turbo-Look coupe	Turbo-Look Targa	Turbo-Look Cabrio
Acceleration	4r	4r	3.5r
Comfort	4.5	4	4.5t
Handling	5	4.5	4
Parts	4	3.5b	4
Reliability	3.5qrs	3.5qrs	3qrst

b - Targa roofs no longer available from Porsche.
q - Alternator, see text.
r - New transmission, clutch, see text.
s - Fuel lines, see text.
t - Power top, improves comfort, risks jamming, see text.

1989 models, manual transmission, continued (5)

	911 Carrera Speedster
Acceleration	3.5r
Comfort	3
Handling	4
Parts	3.5v
Reliability	3.5qrs

q - Alternator, see text.
r - New transmission, clutch, see text.
s - Fuel lines, see text.
v - Speedster fiberglass "humps" very costly to replace.

Bott fitted the car with a manual top, only, that they stored below the humps. It is a tricky operation to raise and lower the top, but with practice it can be done very quickly. For more information, see Chapter 2, devoted entirely to Porsche's open cars.

Then midyear, Porsche introduced model year 1990 Typ 964 Carrera 4 with subtle new styling. The "4" designated full-time all-wheel drive, completing a project Ferdinand Piëch had advocated immediately upon Peter Schutz's arrival at Porsche in 1981. With this car came a new 3.6-liter engine, many changes in the unitized body, and introduction of antilock brakes. This model reenergized the viability of Porsche's 911 model. The 964 was first available only as an all-wheel-drive coupe, but within months Porsche introduced the Cabrio and Targa and offered the three body styles in both Carrera 4 all-wheel- and Carrera 2 rear-wheel-drive models.

This radically new car quickly suffered serious mechanical teething problems, including consequential oil leaks; cylinder-head-to-engine-case leaking problems (Porsche devised an engine without cylinder head gaskets); dual-mass flywheels; and dual-distributor drive belt failures. None of these problems will still exist, all having been remedied through recalls and routine service.

The first year, 1989 1/2, Carrera 4s (and those through the 964 line-up through model year 1993) were somewhat "chunky"

in their operation and generally now are regarded as excellent choices for owners in snowy environs. Later production C4 models and those in the 993 and 996 families are much refined and extremely comfortable as a daily driver in all climates.

1989 Garage Watch
Problems with (and improvements to) Porsche 911 models.

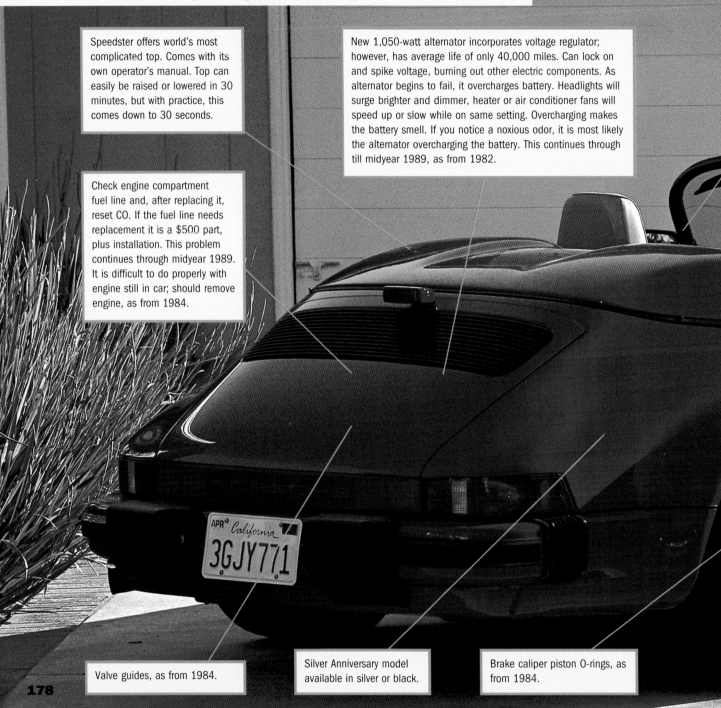

Speedster offers world's most complicated top. Comes with its own operator's manual. Top can easily be raised or lowered in 30 minutes, but with practice, this comes down to 30 seconds.

New 1,050-watt alternator incorporates voltage regulator; however, has average life of only 40,000 miles. Can lock on and spike voltage, burning out other electric components. As alternator begins to fail, it overcharges battery. Headlights will surge brighter and dimmer, heater or air conditioner fans will speed up or slow while on same setting. Overcharging makes the battery smell. If you notice a noxious odor, it is most likely the alternator overcharging the battery. This continues through till midyear 1989, as from 1982.

Check engine compartment fuel line and, after replacing it, reset CO. If the fuel line needs replacement it is a $500 part, plus installation. This problem continues through midyear 1989. It is difficult to do properly with engine still in car; should remove engine, as from 1984.

Valve guides, as from 1984.

Silver Anniversary model available in silver or black.

Brake caliper piston O-rings, as from 1984.

APR California 3GJY771

Convertible tops, as from 1986.

New weather sealing for Targa roofs, a problem since 1967. This is no improvement, but it now incorporates the molding into the rubber, as from 1988.

Heater blowers fail, as from 1975; front fresh air blowers continue to fail, as from 1969.

Brake master cylinders leak. A problem finally remedied in mid-1989 for model year 1990 cars. The brakes would not fully release after being applied. This resulted from pedal bushings stiffening up or being swollen due to rust. The pistons stuck in the calipers. Brake hoses swelled shut, acting like a one-way valve. This problem disappeared with the introduction of power brakes and its resulting complete redesign of the brake system, but all master cylinders can leak after about 10 years, as from 1965.

Age problem appearing. The flexible portions of the fuel lines from the tank into the body—down the tunnel at the center of the car—then out of the tunnel to the engine, are beginning to rot from age and ozone exposure, causing fuel leaks at high pressure, especially in CIS cars, at about 75 psi. There is a fire risk. These leaks can be seen only from underneath the car. If you smell fuel or see a puddle that smells like gas, do not start the car, as from 1973. This is remedied in new type of fuel line fittings on model year 1990 cars, introduced midyear, 1989.

All clutch updates now standard. Many special editions.

Electric seats operation, as from 1985.

1989½-1990

Chapter 30
The New "L" Program—
If You Add 25 Years
to 911, You Get 964

"Remarkably stable, wickedly fast, and astoundingly easy to drive, this 911 comes as a pointed fist in the face of doubters," *Car and Driver* judged in its August 1989 road test. "Here at last is a finely conceived, exquisitely built, ruggedly rocketlike German sports car that redefines the delivery of all-out road performance via four-wheel drive."

While the magazines had test models in late summer 1989, the paying public had to wait until mid–model year. Within six months, Porsche then released the rear-wheel-drive Carrera 2, and now it offered both technologies in coupe, Cabrio, and Targa body styles.

Where the previous Porsche Carrera, the 3.2-liter 1984 model, boasted that it was 80 percent new, with this model, Porsche announced that it had replaced 85 percent of the parts, including some major pieces such as the floor pan and suspension, engine, driveline, and brakes. For model year 1990, Porsche also made driver and passenger air bags standard equipment. (They were not offered on the 1989 1/2 C4 models.)

Porsche began production of the C4 in January 1989 and the C2 models in July. Weissach engineers enlarged the bore from 95 millimeters to 100 and lengthened stroke from 74.4 millimeters to 76.4, creating an overall total displacement of 3,605 cc, or 219.9 cid. The engine, controlled by the Bosch Digital Motor Electronics, or Motronic system, developed 247 SAE net horsepower (and 250 DIN net) at 6,100 rpm and produced 229 SAE net ft-lb of torque (228 DIN net) at 4,800 rpm. The new engine utilized hydraulic timing chain tensioners and ceramic exhaust port liners that reduced heat transfer from the cylinders to the heads.

It was not a perfect system. The engine used twin spark plugs per cylinder, requiring dual distributors that are belt driven. It is easy to determine if the belt has broken. The engine runs horribly because one distributor is firing correctly and the other is firing the plug nearest to the rotor. You have six plugs firing normally and a seventh firing all the time.

From early production 1989 1/2 C4 models through model year 1991, the engines, designed to function without a cylinder head gasket, experienced cylinder head leaks. This was not minor weeping, but serious leaking. The repair required replacing pistons and cylinder barrels, as well as machine work on the crankcase to fit deeper oil skirts. Porsche covered this in many more cars than those few that actually were affected. There were also oil leaks past the through-bolt O-rings and from the timing chain housing. These repairs would cost about $5,000, including parts and labor. It is most likely that this work already has been done on these models. If you are looking at one of these, be sure to ask.

In addition, Porsche mounted the front oil cooler's thermostat housing at the right rear wheel. The housing and the oil lines leading to and from it leaked and failed.

Perhaps the new model's most notable failure was Porsche's new dual-mass flywheel. Weissach created this double-mass flywheel to dampen vibration from the engine that was causing a rattling noise in the Getrag G50 transmission. Freudenberg produced this flywheel that took the rubber elements out of the clutch plate and put them in the flywheel. If the driver slipped the clutch consistently, this overheated the clutch, burned the lining, and caused the disc liner springs to lose tension. Or the rubber damper has broken after getting overheated.

In May 1992, Porsche replaced the Freudenberg flywheels with the "LUK" update, which has the starter ring gear welded to the flywheel. Again, this most likely has been updated under warranty or repaired outside of warranty (about $1,100) in any car you may look at. To be certain, let the clutch out very slowly and listen carefully for a little "clank." (These would indicate loose or fallen starter gear or secondary flywheel mounting bolts, or the heat shield on the secondary flywheel that has come loose or has broken.) If you hear the clank, the Freudenberg dual-mass flywheel has broken. Do not drive this car.

As Porsche models became more electronically intensive and mechanically complex, it's important to understand that if

1989 1/2 Specifications "L" Series

Body Designation:		New Typ 964 Carrera 4
Price:		964 Carrera 4 coupe: $69,500
Engine Displacement and Type:		911 Carrera: Typ 930/25; 3,605 cc, 219.9 cid SOHC, Bosch Digital Motor Electronic (DME) management system
Maximum Horsepower @ rpm:		Carrera 4: 247 SAE net @ 6,100 rpm
Maximum Torque @ rpm:		Carrera 4: 228 ft-lb @ 4,200 rpm
Weight:		Carrera 4: 3,252 pounds
0–60 mph:		Carrera 4: 5.7 seconds (factory)
Maximum Speed:		Carrera 4: 162 mph (factory)
Brakes:		Carrera: power, vented disc brakes w/ABS antilocking brake system
Steering:		hydraulically assisted rack-and-pinion
Suspension:	Front:	MacPherson struts, lower wishbones, coil springs, antisway bar
	Rear:	semi-trailing arms, coil springs, antisway bar
Tires:	Front:	911 Carrera: 205/55ZR16
	Rear:	911 Carrera: 225/50ZR16
Tire air pressure:	Front:	29 psi; Rear: 36 psi
		Typ 964 Front: 36 psi; Rear: 43 psi
Transmission(s):		911 Carrera: G50/01 5-speed U.S.
Wheels:	Front:	911 Carrera: 6.0Jx15 Fuchs alloy
	Rear:	911 Carrera: 8.0Jx16 Fuchs alloy

What they said at the time– Porsche for 1989 1/2

1989 1/2 Porsche Carrera 4, *Car and Driver*, August 1989

"We once wanted to read the minds of the people who create Porsches. Years later, we've learned to do it by driving their cars. . .

"The Carrera 4 is the most technically sophisticated 911 ever to enter series production. A rousing marriage of four-wheel drive, a 3.6-liter, rear-mounted flat six producing 247 horsepower and a stout platform with a redesigned suspension creates a whole as seamless as a ball bearing."

they are running poorly, you want to shut them down as quickly as possible. The problems get more expensive the longer you drive them while they are ailing.

By the same token, as a potential buyer, you don't need to worry about the insides of these engines. They don't lose compression at 150,000 miles. Cars with 80,000 or 90,000 miles are still very strong. You could have a failure if the previous owner ran low on oil, used the wrong heat range of plugs, or consistently lugged the engine, that is, consistently shifted gears below 3,000 rpm. For many first-time Porsche buyers (and the reliability of these vehicles has inspired a number of new owners to buy a 964—or 993 or 996),

running the engine past 4,500 rpm is very difficult. They think they are doing something wrong. Yet the reality is that if they don't run these engines up, they are doing something very harmful.

Porsche largely revised the suspension for the new 964, retaining MacPherson struts in front but trading the long-lived torsion bars at the rear for coil springs and tubular shock absorbers, similar to what's up front. In an obvious attempt to tame the handling once and for all, the engineers adapted the Weissach rear axle from the 928 models with its "track correction" capabilities, mimicking a subtle four-wheel steering. This led to criticism initially that the engineers tamed the rear-engined cars too much.

In fact, with the additional weight up front, the Carrera 4 all-wheel-drive models understeered and plowed in turns, not a previously common criticism for a 911. What the Weissach rear axle did accomplish was to almost completely eliminate the possibility of hanging the rear end out. It took real work. With the C4 models, it required a disastrous screw-up to get the car loose at all.

The Typ 964 was the final design project of Tony Lapine. From the viewpoint of many observers, it improved on Butzi Porsche's original 911 lines without destroying the form. These same critics give Lapine's successor, Harm Lagaay, lower marks for his evolutionary 993 and 996 models that followed. Yet others vigorously disagree, liking Lagaay's lines

strongly. Lapine retained Butzi's stovepipe headlight fenders but rounded things so effectively that he and the aerodynamicists cut the coefficient of drag—Cd—from 0.395 to 0.32, a significant difference that affects wind noise and fuel economy as well as aesthetics. They did this not only by using thermoplastic front and rear bumpers but also by fitting a flat undertray. What no one could understand, however, was why, with this more rounded body, Porsche retained the rectangular, flat-backed outside mirrors. Many buyers later replaced them with the more visually appealing aero-mirrors from Lagaay's 993 when they became available in late 1992 for model year 1993.

1990 Specifications "L" Series

Body Designation:		Typ 964 Carrera 2, 4
Price:		964 Carrera 2 coupe: $58,500
		Carrera Tiptronic: $63,650
		Cabriolet: $66,800
		Targa: $59,900
		964 Carrera 4 coupe: $69,500
		Cabriolet: $66,800
		Targa: $70,900
Engine Displacement and Type:		Carrera 2, 4: Typ M64/01; 3,605 cc, 219.9 cid, SOHC Bosch Digital Motor Electronic (DME)
		management system, dual-spark, twin-plugs
		(Typ 64/02 as above for Tiptronic models)
Maximum Horsepower @ rpm:		Carrera 2, 4: 247 SAE @ 6,100 rpm
Maximum Torque @ rpm:		Carrera 2, 4: 228 ft-lb @ 4,200 rpm
Weight:		Carrera 2 coupe: 3,031 pounds
		Carrera 4 coupe: 3,252 pounds
		Carrera 2: 3,031
		Tiptronic add: 56 pounds
0–60 mph:		Carrera 2: 5.5 seconds (factory)
		Carrera 4: 5.7 seconds (factory)
		Carrera 2/Tiptronic: 6.4 seconds (factory)
Maximum Speed:		Carrera 2, 4: 162 mph (factory)
		Carrera 2 Tip: 161 mph (factory)
Brakes:		Carrera: power vented disc brakes w/ABS anti-locking brake system
Steering:		hydraulically assisted rack-and-pinion
Suspension:	Front:	MacPherson struts with telescoping shock absorbers, lower control arms
	Rear:	semi-trailing arms, coil springs with toe correction
Tires:	Front:	Carrera: 205/55ZR16
	Rear:	Carrera: 225/50ZR16
Tire air pressure:	Front:	36 psi; Rear: 43 psi
Transmission(s):		Carrera 4: G64/00 5-speed
		Carrera 2: G50/03 5-speed
		Tiptronic: G50/01
Wheels:	Front:	Carrera: 6.0Jx16 Fuchs alloy
	Rear:	Carrera: 8.0Jx16 Fuchs alloy

What they said at the time–Porsche for 1990

911 Carrera 2, *Autosport*, January 1990

"Arguably, the biggest change previous 911 owners will notice is that the clumsy, overcenter clutch now works with impeccable smoothness and weighting, and the gear change has been improved out of all recognition."

Carrera 2 Tiptronic, *Car South Africa*, September 1990

"Although this is a proper torque converter transmission, adding some 66 pounds to the car's weight and costing about one second in 0–100 times and 2.5 miles per hour in top speed, it avoids both ratio changes in a corner (by reading lateral acceleration) and overrun upshifts when you lift off before a bend. It also senses braking, to avoid sudden power changes on slippery surfaces, while raising idle revs when the engine is cold."

What they said at the time continued

Carrera 4, *Road & Track*, March 1990

"Of all the components you don't readily see, none is more important than the Carrera 4's all-wheel drive system. A center differential, mounted at the leading edge of a new five-speed gearbox, uses a planetary gear and an electronically controlled multidisc clutch to send 31 percent of the engine's output to the front wheels and 69 percent to the rear. But when the car's wheel-mounted ABS sensors detect slippage at any corner, power is reapportioned so that optimal traction is maintained at all times."

Parts List for 1990 911s (964)

These are items most commonly replaced during regular maintenance and routine daily operation. Prices quoted are for new factory parts at list price, not including installation labor. NLA means factory parts no longer available, so prices quoted are from aftermarket suppliers.

Engine:

1. Oil filter..................... $15.82
2. Alternator belt set..... $57.14
3. Starter....................... $730.24
4. Alternator................. $942.86
5. Muffler..................... $1,702.05
6. Clutch disc............... $284.56

Body:

7. Front bumper............. $1,657.70
8. Left front fender........ $1,780.45
9. Right rear quarter panel............. $2,088.84
10. Front deck lid........... $1,797.09
11. Front deck lid struts.. $37.23 each
12. Rear deck lid struts... $32.88 each
13. Porsche badge, front deck lid............. $149.76
14. Taillight housing and lens.................... $348.13
15. Windshield with antenna............. $1,534.34
16. Windshield weather stripping................... $200.54

Interior:

17. Dashboard................. $2,863.88
18. Shift knob................. $237.07
19. Interior carpet, complete................... $900.00

Chassis:

20. Front rotor................. $300.76
21. Brake pads, front set (C2)........... $208.69
22. Koni rear shock absorber.......... $249.924
23. Front wheel.............. $910.61
24. Rear wheel............... $1,007.92

Ratings

1990 models, manual transmission

	911 Carrera 2 coupe	Carrera 2 Targa	Carrera 2 Cabrio
Acceleration	4	4	4
Comfort	4	4	4
Handling	4.5	4	4
Parts	3.5	3.5b	3.5
Reliability	2.5w	2.5w	2.5w

b - Targa roofs no longer available from Porsche.
w - Numerous serious new-model teething problems, see text.

1990 models, manual transmission, continued

	911 Carrera 4 coupe	Carrera 4 Targa	Carrera 4 Cabrio
Acceleration	4	4	4
Comfort	4	4	4
Handling	5	4.5	4
Parts	3.5	3.5b	3.5
Reliability	2.5w	2.5w	2.5w

b - Targa roofs no longer available from Porsche.
w - Numerous new-model teething problems, see text.

1990 models, Tiptronic transmission

	911 Carrera 2 coupe	Carrera 2 Targa	Carrera 2 Cabrio
Acceleration	3.5	3.5	3.5
Comfort	4	4	4
Handling	4	4	4
Parts	3.5	3.5b	3.5
Reliability	2.5w	2.5w	2.5w

b - Targa roofs no longer available from Porsche.
w - Numerous new-model teething problems, see text.

1990 Garage Watch
Problems with (and improvements to) Porsche 911 models.

Windshield and rear windows leak in corners when rubber seals get old and brittle, as from 1965.

Carrera 2, rear-wheel-drive Typ 964 introduced.

Dashboard switches on C4 to engage locking rear differential and other to raise rear spoiler manually.

Oil cooler thermostat housing and lines leak, fail.

Last year cast aluminum intake manifold.

Air conditioning system for 1989 C4, and 1990 C2 and C4 is R-12 system, with fragile front condenser and evaporator coils. They often leak, but the systems are easily converted to 134A. Also there are problems with electric controls for the AC. These are complicated, troublesome, and expensive to repair.

Fuel tank sending units recalled.

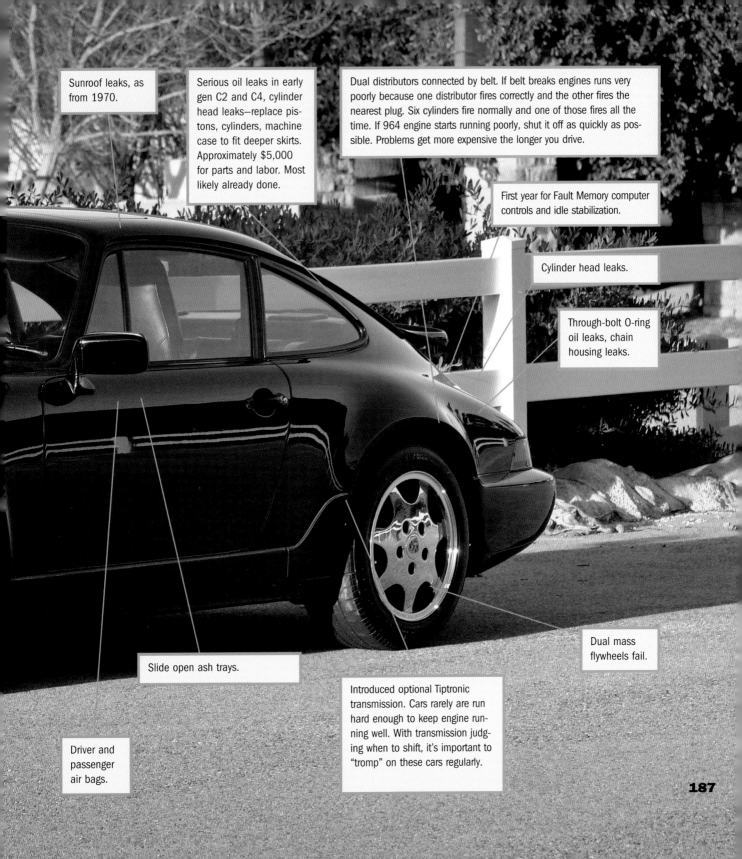

Sunroof leaks, as from 1970.

Serious oil leaks in early gen C2 and C4, cylinder head leaks—replace pistons, cylinders, machine case to fit deeper skirts. Approximately $5,000 for parts and labor. Most likely already done.

Dual distributors connected by belt. If belt breaks engines runs very poorly because one distributor fires correctly and the other fires the nearest plug. Six cylinders fire normally and one of those fires all the time. If 964 engine starts running poorly, shut it off as quickly as possible. Problems get more expensive the longer you drive.

First year for Fault Memory computer controls and idle stabilization.

Cylinder head leaks.

Through-bolt O-ring oil leaks, chain housing leaks.

Slide open ash trays.

Dual mass flywheels fail.

Introduced optional Tiptronic transmission. Cars rarely are run hard enough to keep engine running well. With transmission judging when to shift, it's important to "tromp" on these cars regularly.

Driver and passenger air bags.

Porsche introduced the 964-bodied Turbo in September 1990 as a 1991 model, using the K-Jetronic-controlled engine with a larger turbo and intercooler and improved exhaust breathing through the catalytic converters standard on C2s and C4s. The turbo boasted 315 horsepower SAE net, 320 horsepower DIN net, at 5,750 rpm and 332 ft-lb SAE and DIN net torque at 4,500. Despite the impact of the all-wheel-drive, six-speed Typ 959, the new car was five-speed rear-wheel drive only.

The normally aspirated all-wheel-drive Carrera 4, introduced mid–model year 1989, and the rear-drive Carrera 2, available model year 1990, remained unchanged. Overall displacement was increased by enlarging the bore and lengthening the stroke. The 3,605cc, 219.9 cid engine developed 247 SAE net horsepower (and 250 DIN net) at 6,100 rpm, and 229 SAE net ft-lb of torque, (228 DIN net) at 4,800 rpm. Hydraulic timing chain tensioners and ceramic exhaust port liners reduced heat transfer from the cylinders to the heads.

It was not perfect. The engine used belt-driven dual distributors to fire two spark plugs per cylinder. If the belt breaks, the engine runs horribly.

From early production 1989 1/2 C4s through 1991 models, the engines, designed to function without a cylinder head gasket, experienced serious cylinder head leaks. Porsche replaced pistons and cylinder barrels and covered machine work on the crankcase to fit deeper oil skirts. There were also oil leaks past the through-bolt O-rings and from the timing chain housing.

In addition, the housing and the oil lines leading to and from the front oil cooler's thermostat, housed at the right rear wheel, often leaked and failed.

Perhaps the new model's most notable failure was Porsche's dual-mass flywheel, introduced in 1989 1/2 on the C4 to dampen engine vibration that caused a rattle in the Getrag G50 transmission. However, if the driver slipped the clutch consistently, the clutch overheated and either burned the lining or broke the rubber damper.

In May 1992, Porsche replaced the Freudenberg flywheels with the "LUK" update. Again, this most likely has been updated (about $1,100) in any car you look at. Let the clutch out very slowly and listen for a "clank." (This indicates loose or fallen starter gear or secondary flywheel mounting bolts, or the heatshield on the secondary flywheel has come loose or broken.) Do not drive this car.

In January 1990, Porsche introduced its Tiptronic transmission, a computer-controlled four-speed that could be operated automatically or shifted manually. Some seals leak. Also, in fully automatic operation, the engine does not get worked hard enough.

Option M573, automatic climate control, introduced in 1989 1/2, brought troublesome and expensive problems.

Beginning with the midyear 1989 Typ 964s, Porsches are more electronically intensive and mechanically complex; if they are running poorly, you must shut them down as quickly as possible. The problems get more expensive the longer you drive them.

By the same token, cars with 80,000 to 90,000 miles are still very strong. You could have a failure if the previous owner ran low on oil, used the wrong plugs, or consistently shifted gears below 3,000 rpm.

For many first-time Porsche buyers (the reliability of these vehicles has inspired a number of new owners to buy a 964—or 993 or 996), getting them to run the engine past 4,500 rpm is difficult. The truth is if you don't run these engines, you are doing something very harmful.

Porsche largely revised the suspension for the new 964, retaining MacPherson front struts but replacing the rear torsion bars with coil springs and shock absorbers. In an obvious attempt to improve the handling, engineers adapted the 928's Weissach rear axle with its "track correction" capabilities. This led to criticism that the engineers had tamed the rear-engined cars. In fact, with the additional weight up front, all-wheel-drive Carrera 4s understeered and plowed in turns.

The Typ 964 was the final design by Tony Lapine. From the viewpoint of many, it improved Butzi Porsche's original 911 lines without destroying the form. Lapine retained Butzi's stovepipe headlight fenders but rounded things so effectively as to cut the Coefficient of drag, favorably affecting wind noise, fuel economy, and aesthetics.

1991 Specifications "M" Series

Body Designation:		Typ 964 Carrera 2, Carrera 4, 911 Turbo 3.3
Price:		964 Carrera 2 coupe: $61,915
		Cabriolet: $70,690
		Targa: $63,445
		964 Carrera 4 coupe: $73,440
		Cabriolet: $82,215
		Targa: $74,970
		Turbo: $95,000
Engine Displacement and Type:		Carrera 4, 2: Typ M64/01; 3,605 cc, 219.9 cid, SOHC, Bosch Digital Motor Electronic (DME) management system, dual-spark, twin-plugs (M64/02, as above, for Tiptronic models).
		Turbo: Typ M30/69; 3,299 cc 201.3 cid, SOHC Bosch DME, Turbocharger, intercooler, 3-way catalytic converters
Maximum Horsepower @ rpm:		Carrera 4, 2: 247 SAE @ 6,100 rpm
		Turbo: 315 SAE @ 5,750
Maximum Torque @ rpm:		Carrera 4, 2: 228 ft-lb @ 4,200 rpm
		Turbo: 332 @ 4,500
Weight:		Carrera 2 coupe: 3,031 pounds
		Carrera 4 coupe: 3,252 pounds add 69 pounds for Tiptronic
		Turbo: 3,274 pounds
0–60 mph:		Carrera 2, 4: 5.5 seconds (factory)
		Carrera Tiptronic: 6.4 seconds (factory)
		Turbo: 4.8 seconds (MT)
Maximum Speed:		Carrera 2, 4: 162 mph (factory)
		Carrera Tip: 159 mph (factory)
		Turbo: 168 mph (factory)
Brakes:		Carrera: power vented disk brakes w/ABS anti-locking brake system
		Turbo: power ventilated disk brakes, 4-piston calipers front, 2-piston calipers rear, ABS,
Steering:		hydraulically assisted rack-and-pinion
Suspension:	Front:	MacPherson struts with telescoping shock absorbers, lower control arms
	Rear:	semi-trailing arms, coil springs with toe correction
Tires:	Front:	Carrera: 205/55ZR16
		Turbo: 205/50ZR17
	Rear:	Carrera: 225/50ZR16
		Turbo: 255/40ZR17
Tire air pressure:		(Carrera and Turbo) Front: 36 psi; Rear: 43 psi
Transmission(s):		Carrera 4: G64/00 5-speed
		Carrera 2: G50/03 5-speed
		Tiptronic: G50/01
		Turbo: G50/52 5-speed w/ZF mechanical limited-slip differential, 20 percent lockup
Wheels:	Front:	Carrera: 6.0Jx16
		Turbo: 7.0Jx17
	Rear:	Carrera: 8.0Jx16
		Turbo: 9.0Jx17

What they said at the time–Porsche for 1991

Carrera 2 Tiptronic, *Car and Driver*, March 1991

"If you leave the selector in "D," while driving hard on twisty roads, you have to be wary of sharp turns connected by short straight-aways on which you're inclined to hold a steady throttle. Looking at your steady right foot, the Bosch computer says, 'Hey, this guy is going for some interstate cruising,' and the transmission responds with a premature upshift from second to third—sometimes even to fourth. This is unnerving when you're briskly bearing down on a second gear corner. (They are virtually all second gear corners; second gear provides flexible power from 38 to 80 miles per hour.)"

Parts List for 1991 911s

These are items most commonly replaced during regular maintenance and routine daily operation. Prices quoted are for new factory parts at list price, not including installation labor. NLA means factory parts no longer available, so prices quoted are from aftermarket suppliers.

Engine:

1. Oil filter.................... $15.82
2. Alternator belt set..... $57.14
3. Starter...................... $730.24
4. Alternator $942.86
5. Muffler...................... $1,702.05
   (Turbo) $1,659.47
6. Clutch disc $284.56

Body:

7. Front bumper............. $1,657.70
8. Left front fender........ $1,780.45
   (Turbo fender
   plus flare) $2,328.77
9. Right rear
   quarter panel............. $2,088.84
10. Front deck lid $1,797.09
11. Front deck lid struts.. $37.23 each
12. Rear deck lid struts... $32.88 each
13. Porsche badge,
    front deck lid............. $149.76
14. Taillight housing
    and lens $348.13
15. Windshield
    with antenna $1,534.34
16. Windshield weather
    stripping.................... $200.54

Interior:

17. Dashboard................. $2,863.88
18. Shift knob $237.07
19. Interior carpet,
    complete $900.00

Chassis:

20. Front rotor................. $300.76
21. Brake pads,
    front set $208.69
22. Koni rear
    shock absorber.......... $249.21
23. Front wheel $910.61
24. Rear wheel $1,007.92

Ratings

1991 models, manual transmission

	Carrera 2 coupe	Carrera 2 Targa	Carrera 2 Cabrio
Acceleration	4	4	4
Comfort	4	4	4
Handling	4.5	4	4
Parts	4	3.5b	4
Reliability	2.5w	2.5w	2.5w

b - Targa roofs no longer available from Porsche.
w - Numerous new model teething problems, see text.

1991 models, manual transmission, continued

	911 Carrera 4 coupe	Carrera 4 Targa	Carrera 4 Cabrio
Acceleration	4	4	4
Comfort	4	4	4
Handling	5	4.5	4
Parts	4	3.5b	4
Reliability	2.5w	2.5w	2.5w

b - Targa roofs no longer available from Porsche.
w - Numerous new model teething problems, see text.

1991 model, manual transmission

	911 Turbo coupe
Acceleration	5
Comfort	5
Handling	4.5
Parts	4
Reliability	3w

w - Numerous new model teething problems, see text.

1991 models, Tiptronic transmission

	911 Carrera 2 coupe	Carrera 2 Targa	Carrera 2 Cabrio
Acceleration	3.5	3.5	3.5
Comfort	4	4	4
Handling	4	4	4
Parts	4	3.5b	4
Reliability	2.5w	2.5w	2.5w

b - Targa roofs no longer available from Porsche.
w - Numerous new model teething problems, see text.

1991 Garage Watch
Problems with (and improvements to) Porsche 911 models.

Ash trays continue to slide open, as from 1989.

Windshield and rear windows leak in corners when rubber seals get old and brittle, as from 1965.

Air conditioning system for 1989 C4, and 1990 and 1991 C2 and C4 are R-12 system, with fragile front condenser and evaporator coils. They often leak, but the systems are easily converted to 134A.

Fuel tank sending units recalled, as from 1989, problem fixed.

Oil cooler thermostat housing and lines leak and fail, as from 1989.

Turbo reintroduced.

Sunroof leaks, as from 1970.

Fault memory computer controls require Bosch computer tool for repairs.

Serious oil leaks in C2 and C4 to midyear. To repair cylinder head leaks, replace pistons, cylinders, machine case to fit deeper skirts. Approximately $5,000 for parts and labor. Most likely already done.

New glass-reinforced palstic (GRP) intake manifold.

Cylinder head leaks—pistons, cylinders replaced.

Dual distributors connected by belt. If belt breaks, engine runs very poorly because one distributor fires correctly and the other fires the nearest plug. Six cylinders fire normally and one fires all the time. If 964 engine starts running poorly, shut it off as quickly as possible. Problems get more expensive the longer you drive, as from 1989 1/2.

Tiptronic seals, as from 1990.

Through-bolt O-ring oil leaks, chain housing leaks.

Dual mass flywheels fail, as from 1989.

Porsche reintroduced the rear-wheel-drive Turbo for 1991, using a larger turbo and intercooler, and the metallic catalytic converters standard on C2s and C4s. For 1992, Porsche offered a limited edition (80 units), Turbo S, labeled the "S2" in America. For reasons never explained, it underrated power horsepower. But the S2's long-legged gearing made for slower acceleration, frustrating the few U.S. customers who came up with the $10,065 optional charge and $7,170 luxury tax on top of the $108,870 list price! The S and S2 went out of production before the end of the model year.

Throughout this model year, European customers could buy the semicompetition Carrera 2 RS inspired by the 1973 Carrera 2.7 RS. However, this engine was neither smog-legal nor DOT-approved in the Untied States. To placate U.S. customers, the factory created the RS America. It was 270 pounds heavier than the European version and just 298 were built. Its most distinctive feature is its fixed whale tail rear wing.

Porsche had introduced the all-wheel-drive Carrera 4 in mid–model year 1989 and the C2 rear drive appeared in model year 1990. Weissach increased displacement by enlarging the bore and lengthening the stroke. Porsche controlled the 3,605 cc engine with the Bosch Digital Motor Electronics (Motronic) system. The engine utilized hydraulic timing chain tensioners and ceramic exhaust port liners to reduce heat transfer.

Belts often broke on the belt-driven dual distributors, causing one distributor to fire properly and the other to spark each time the plug nearest to where the rotor stuck.

Perhaps the new model's most notable failure was the dual-mass flywheel, introduced on the 1989 1/2 C4. Created to dampen engine vibration that caused a rattle in the transmission, the flywheel and clutch overheated if the driver slipped the clutch consistently. There are also reports that the clutch disc liner springs lost tension.

The troublesome flywheels also incorporated bolted-on starter ring gears that worked loose. Near the end of model year 1992, Porsche replaced the flywheels with the "LUK" update, which welded the starter ring gear to the flywheel. This most likely has been updated (about $1,100) in any car you look at. To check, let the clutch out slowly and listen for a "clank" that indicates loose or fallen starter gear or secondary flywheel mounting bolts, or a loose or broken heatshield. Do not drive this car.

In January 1990, Porsche introduced its computer-controlled four-speed Tiptronic transmission that could be operated fully automatically or shifted manually. Some seals leak. Also, in fully automatic operation, the engine does not get worked hard enough.

Option M573, automatic climate control, introduced in 1989 1/2, brought complicated, troublesome, and expensive problems.

Now that Porsches are more electronically and mechanically complex, if they run poorly, you must shut them down ASAP. The problems get more expensive the longer you drive them.

By the same token, there are fewer worries about the inside of these engines. They don't lose compression at 150,000 miles. You could have a failure if the previous owner ran low on oil, used the wrong plugs, or consistently shifted gears below 3,000 rpm.

The improved reliability of 964, 993, and 996 has enticed many individuals to buy their first Porsche. But getting first-time owners to run the engine past 4,500 rpm is difficult. The truth is, it's more harmful if you don't run these engines.

For the new 964, Porsche retained MacPherson front struts but replaced the rear torsion bars with coil springs and shock absorbers. In an obvious attempt to tame the handling, engineers adapted the 928's Weissach rear axle with its "track correction" capabilities, leading to criticism that they had tamed the cars. In fact, with the additional weight up front, all-wheel-drive Carrera 4s understeered and plowed in turns.

The Typ 964 was Tony Lapine's final project. Many felt it improved Butzi Porsche's original 911 lines. Lapine retained Butzi's stovepipe headlight fenders but rounded things so effectively as to cut the coefficient of drag, favorably affecting wind noise, fuel economy, and aesthetics. No one could understand, however, why Porsche retained the rectangular outside mirrors, which many buyers replaced with aero-mirrors from Lagaay's 993, which became available for model year 1995.

The front-engine, water-cooled 944S evolved into the Typ 968 for 1992, supplemented by the new, luxurious 928GTS.

1992 Specifications "N" Series

Body Designation:	Typ 964 Carrera 4, Carrera 2, America Roadster, Turbo 3.3
Price:	964 Carrera 2 coupe: $63,900
	Cabriolet: $72,900
	Targa: $65,500
	America Roadster: $87,900
	964 Carrera 4 coupe: $75,780
	Cabriolet: $84,780
	Targa: $77,380
	Turbo: $98,875
	Turbo S: $118,935
Engine Displacement and Type:	Carrera 4, 2, RS America: Typ M64/01; 3,605 cc, 219.9 cid, SOHC, Bosch Digital Motor Electronic (DME) management system, dual-spark, twin-plugs (M64/02, as above, for Tiptronic models.)
	Turbo: Typ M30/69; 3,299 cc 201.3 cid, SOHC, Bosch DME, Turbocharger, intercooler, 3-way catalytic converters
	Turbo S: M30/69SL; 3,299 cc, 201.3 cid, SOHC, remapped Bosch DME, Turbocharger, intercooler, 3-way catalytic converters
Maximum Horsepower @ rpm:	Carrera 2, 4: 247 SAE @ 6,100 rpm
	Turbo: 315 SAE @ 5,750
	Turbo S: 322 SAE @ 6,200 rpm
Maximum Torque @ rpm:	Carrera 2, 4: 228 ft-lb @ 4,800 rpm
	Turbo: 332 @ 4,500
	Turbo S: 370 @ 4,800
Weight:	Carrera 2, 4 coupe: 3,252 pounds, add 66 pounds for Tiptronic
	RS America: 2,954 pounds
	Turbo: 3,274 pounds
	Turbo S: 2,844 pounds
0–60 mph:	Carrera 2, 4, and America Roadster: 5.5 seconds (factory)
	Carrera Tiptronic: 6.4 seconds (factory)
	Turbo: 4.8 seconds (factory)
	Turbo S: 4.9 seconds (factory)
Maximum Speed:	Carrera 4, 2: 162 mph
	Carrera Tip: 159 mph
	Turbo: 168 mph (factory)
	Turbo S: 178 mph (factory)
Brakes:	Carrera: power vented disk brakes w/ABS antilocking brake system
	RS America, Turbo, and Turbo S: vented, cross-drilled hydraulically assisted, dual circuit, four-piston fixed calipers, ABS
Steering:	Hydraulically assisted rack-and pinion
Suspension: Front:	MacPherson struts with telescoping shock absorbers, coil springs, lower control arms, antisway bar
Rear:	semi-trailing arms, coil springs, self-stabilizing toe correction, antisway bar
Tires: Front:	Carrera: 205/55ZR16
	RS America, Turbo: 205/50ZR17
	Turbo S: 235/40ZR18
Rear:	Carrera: 225/50ZR16
	RS America, Turbo: 255/45ZR17
	Turbo S: 265/35ZR19
Tire air pressure:	All models: Front and rear: 36 psi
Transmission(s):	Carrera 4: G64/00 5-speed
	Carrera 2: G50/03 5-speed
	Tiptronic: A50/03
	Turbo and S: G50/52 5-speed w/ZF mechanical limited-slip differential, 20 percent lock up

1992 Specifications "N" Series continued

Wheels:	Front:	Carrera: 6.0Jx15
		RS America, Turbo: 7.0Jx17
		Turbo S: 8.0J-18
	Rear:	Carrera: 8.0Jx16
		RS America, Turbo: 9.0Jx17
		Turbo S: 10.0J-18

What they said at the time–Porsche for 1992

1992 Carrera 2 Cabrio Tiptronic, *Autocar*, January 1992

"You aren't allowed to reach the redline in first. As a precaution, the Tiptronic is programmed to move into second well short of the 6,800 rpm redline. In itself that's only marginally annoying, but it often won't change down again when you want to, with the consequence that you flounder out of hairpins at low revs in second gear, well below the torque peak (219 ft-lb at 4,800 rpm). . . You soon feel frustrated. Not only has Porsche provided no more than four gears, but its engineers apparently do not trust you sufficiently to let you use all of them to the full."

Parts List for 1992 911s

These are items most commonly replaced during regular maintenance and routine daily operation. Prices quoted are for new factory parts at list price, not including installation labor. NLA means factory parts no longer available, so prices quoted are from aftermarket suppliers.

Engine:

1. Oil filter $15.82
2. Alternator belt set $57.14
3. Starter $730.24
4. Alternator $942.86
5. Muffler $1,702.05
   (Turbo) $1,659.47
6. Clutch disc $284.56

Body:

7. Front bumper $1,657.70
8. Left front fender $1,780.45
   (Turbo fender
   plus flare) $2,328.77
9. Right rear
   quarter panel $2,088.84
10. Front deck lid $1,797.09
11. Front deck lid struts .. $37.23 each
12. Rear deck lid struts ... $32.88 each
13. Porsche badge,
    front deck lid $149.76
14. Taillight housing
    and lens $348.13
15. Windshield
    with antenna $1,534.34
16. Windshield weather
    stripping $200.54

Interior:

17. Dashboard $2,863.88
18. Shift knob $237.07
19. Interior carpet,
    complete $900.00

Chassis:

20. Front rotor $300.76
21. Brake pads,
    front set $208.69
22. Koni rear
    shock absorber $249.24
23. Front wheel $910.61
24. Rear wheel $1,007.92

Ratings

1992 models, manual transmission

	Carrera 2 coupe	Carrera 2 Targa	Carrera 2 Cabrio
Acceleration	4	4	4
Comfort	4.5	4.5	4.5
Handling	4.5	4	4
Parts	4	3.5b	4
Reliability	3w	3w	3w

b - Targa roofs no longer available from Porsche.
w - Lingering teething problems, see text.

1992 models, manual transmission, continued (2)

	911 Carrera 4 coupe	Carrera 4 Targa	Carrera 4 Cabrio
Acceleration	4	4	4
Comfort	4.5	4.5	4.5
Handling	5	4.5	4
Parts	4	3.5b	4
Reliability	3w	3w	3w

b - Targa roofs no longer available from Porsche.
w - Lingering teething problems, see text.

Ratings

1992 model, manual transmission, continued (3)

	911 Turbo coupe	911 RS America coupe
Acceleration	5	4
Comfort	5	2.5
Handling	4.5	4.5
Parts	4	4
Reliability	3w	3w

w - Lingering teething problems, see text.

1992 models, Tiptronic transmission

	911 Carrera 2 coupe	Carrera 2 Targa	Carrera 2 Cabrio
Acceleration	3.5	3.5	3.5
Comfort	4.5	4.5	4.5
Handling	4.5	4	4
Parts	4	3.5b	4
Reliability	3w	3w	3w

b - Targa roofs no longer available from Porsche.

w - Lingering teething problems, see text.

1992 Garage Watch
Problems with (and improvements to) Porsche 911 models.

RS America reintroduced.

Slide open ash trays, as from 1989, now fixed.

Windshield and rear windows leak in corners when rubber seals get old and brittle, as from 1965.

Sunroof leaks, as from 1970.

Tiptronic seals, as from 1990.

3.3-liter Turbo ends.

Thermostat housing, oil lines, as from 1989.

O-ring oil leaks, as from 1989.

Dual mass flywheel still fails, as from 1989.

Cylinder head leaks, as from 1989.

Dual distributors connected by belt. If belt breaks engines runs very poorly because one distributor fires correctly and the other fires the nearest plug. Six cylinders fire normally and one sparks all the time. If 964 engine starts running poorly, shut it off as quickly as possible. Problems get more expensive the longer you drive, as from 1989 1/2.

Four-piston rear brake calipers introduced on C2, with proportioning valve to prevent lockup.

199

Porsche waited until mid–model year 1993 to introduce the Turbo 3.6 in the U.S. customers. Technically a 1994, the car carried over the 3.3-liter model's suspension, but was lowered 20 millimeters and included a front shock tower crossbrace. The new version was a substantially more powerful car, increasing output to 355 horsepower and pumping up torque to 384 ft-lb SAE net. Porsche retained the larger turbo and intercooler of the 3.3 and the metallic catalytic converters of the C2 and C4.

Porsche had introduced the all-wheel-drive Carrera 4 in mid–model year 1989 and the C2 rear-drive appeared in model year 1990. Engineers increased overall engine displacement by enlarging the bore and lengthening the stroke. The 3,605 cc engine was controlled with the Bosch Digital Motor Electronics (Motronic) system. The engine developed 247 SAE net horsepower and produced 229 SAE net ft-lb of torque. It utilized hydraulic timing chain tensioners and ceramic exhaust port liners.

Belts often broke on the belt-driven dual distributors, causing one distributor to fire properly and the other to spark the same plug repeatedly.

Perhaps the new model's most notable failure was the dual-mass flywheel, introduced on the 1989 1/2 C4. Created to dampen engine vibration that caused a rattle in the transmission, the flywheel and clutch overheated if the driver slipped the clutch consistently. There are also reports that the clutch disc liner springs lost tension.

These flywheels incorporated bolted-on starter ring gears that sometimes worked loose. Near the end of model year 1992, Porsche replaced the flywheels with the "LUK" update. This most likely has been updated (about $1,100) in any car you look at. To check, let the clutch out slowly and listen for a "clank" that indicates loose or fallen starter gear or secondary flywheel mounting bolts, or a loose or broken heatshield. Do not drive this car.

In January 1990, Porsche introduced its computer-controlled four-speed Tiptronic transmission that could be operated fully automatically or shifted manually. Some seals leak. Also, in fully automatic operation, the engine does not get worked hard enough.

Now that Porsche models are more electronically intensive and mechanically complex, if they are running poorly, you must shut them down as soon as possible. The problems get more expensive the longer you drive them.

By the same token, as a potential buyer, you have fewer worries about the insides of these engines. They don't lose compression at 150,000 miles. You could have a failure, however, if the previous owner ran low on oil, used the wrong plugs, or consistently shifted gears below 3,000 rpm.

The improved reliability of 964, 993, and 996 models has enticed many individuals to buy their first Porsche. But getting first-time owners to run the engine past 4,500 rpm is difficult. The truth is that if you don't run these engines, you are doing more harm than good.

For the new 964, Porsche retained MacPherson front struts but replaced the rear torsion bars with coil springs and shock absorbers. In an obvious attempt to tame the handling, engineers adapted the 928's rear axle, leading to cries that the cars were over-tamed. In fact, with the additional weight up front, all-wheel-drive Carrera 4s understeered and plowed in turns.

Responding to environmental concerns, Porsche switched its air conditioning refrigerant from Freon-type CFC coolant to HFC. Some seals on the new system were vulnerable to leakage. In addition, option M573 automatic climate control, introduced in 1989 1/2, brought complicated, troublesome, and expensive problems.

The Typ 964 was Tony Lapine's final design. In 1993, Porsche offered several models based on the C2, including the RS America and a new America Roadster. The Typ 964 body lent itself to a U.S. customer base hungry for distinctive variations. The America Roadster used the Turbo cabriolet body with the Carrera 2 3.6-liter normally aspirated engine. It deleted rear seats but came with the power lift top. Porsche produced only 250.

Lapine retained Butzi Porsche's stovepipe headlight fenders while cutting the coefficient of drag, improving wind noise, fuel economy, and aesthetics. No one could understand, however, why Porsche retained the rectangular outside mirrors, which many buyers replaced with aero-mirrors from Lagaay's 993, which became available for model 1995.

1993 Specifications "P" Series

Body Designation:		Typ 964 Carrera 2, Cabriolet, Targa, America Roadster, RS America
Price:		964 Carrera 2 coupe: $64,990
		RS America: $54,800
		Cabriolet: $74,190
		Targa: $66,600
		America Roadster: $89,350
		Carrera 4 coupe: $78,450
		Turbo 3.6: $99,000
Engine Displacement and Type:		Carrera 2, 4: Typ M64/01; 3,605 cc, 219.9 cid, SOHC, Bosch Digital Motor Electronic (DME) management system, dual-spark, twin-plugs (M64/02, as above, for Tiptronic models)
Maximum Horsepower @ rpm:		Carrera 2, 4: 247 SAE @ 6,100 rpm
Maximum Torque @ rpm:		Carrera 2, 4: 228 ft-lb @ 4,200 rpm
Weight:		Carrera 2 coupe: 3,031 pounds
		RS America: 2,954 pounds
		America Roadster: 3,164 pounds
		Carrera 4 coupe: 3,362 pounds, add 56 pounds for Tiptronic
		Turbo: 3,274 pounds
0–60 mph:		Carrera 2, American Roadster: 5.5 seconds (factory)
		RS America: 5.4 seconds (factory)
		Carrera 4: 5.5 seconds (factory)
		Carrera Tiptronic: 6.4 seconds (factory)
		Turbo: 4.7 seconds (factory)
Maximum Speed:		Carrera 2, RSA: 162 mph (factory)
		Carrera Tip: 159 mph (factory)
		America Roadster, C4 coupe: 158 mph (factory)
		Turbo: 174 mph (factory)
Brakes:		Carrera and Turbo: power vented disc brakes w/ABS anti-locking brake system
Steering:		hydraulically assisted rack-and-pinion
Suspension:	Front:	MacPherson struts with telescoping shock absorbers, lower control arms
	Rear:	semi-trailing arms, coil springs with toe correction
Tires:	Front:	Carrera: 205/55ZR16
		Turbo: 225/40ZR18
	Rear:	Carrera: 225/50ZR16
		Turbo: 255/45ZR17
Transmission(s):		Carrera 4: G64/00 5-speed
		Carrera 2: G50/03 5-speed
		Tiptronic: A50/03
		Turbo: G50/52 5-speed w/ZF mechanical limited-slip differential, 20 percent lockup
Wheels	Front:	Carrera: 6.0Jx15
		Turbo: 8.0Jx18
	Rear:	Carrera: 8.0Jx16
		Turbo: 10.0Jx18

What they said at the time–Porsche for 1993

911 Turbo S2, *Car and Driver*, February 1993

"In fact, what we have here is the fastest street-going model that Porsche has ever sold in America. We measured a top speed of 178 miles per hour, up from the standard Turbo's 166 miles per hour. And the S2 engine pulled strongly up to its 6,600 redline in every gear but fifth—and almost did that too.

"Truth is, if this engine is making in the neighborhood of 370 horsepower, then we've all been giving Sir Isaac Newton and his laws of motion way too much credit for the past several centuries."

Parts List for 1993 911s

These are items most commonly replaced during regular maintenance and routine daily operation. Prices quoted are for new factory parts at list price, not including installation labor. NLA means factory parts no longer available, so prices quoted are from aftermarket suppliers.

Engine:

1. Oil filter.................... $15.82
2. Alternator belt set..... $57.14
3. Starter...................... $730.24
4. Alternator................. $942.86
5. Muffler $1,702.05
   (Turbo) $1,659.47
6. Clutch disc............... $284.56

Body:

7. Front bumper............. $1,657.70
8. Left front fender........ $1,780.45
9. Turbo fender
   plus flare.................. $2,328.77
10. Right rear
    quarter panel............ $2,088.84
11. Front deck lid........... $1,797.09
12. Front deck lid struts.. $37.23 each
13. Rear deck lid struts... $32.88 each
14. Porsche badge,
    front deck lid............ $149.76
15. Taillight housing
    and lens $348.13
16. Windshield
    with antenna $1,534.34
17. Windshield weather
    stripping.................... $200.54

Interior:

18. Dashboard................. $2,863.88
19. Shift knob $237.07
20. Interior carpet,
    complete $900.00

Chassis:

21. Front rotor................. $300.76
22. Brake pads,
    front set $208.69
23. Koni rear
    shock absorber.......... $249.24
24. Front wheel $910.61
25. Rear wheel $1,007.92

Ratings

1993 models, manual transmission

	Carrera 2 coupe	Carrera 2 Targa	Carrera 2 Cabrio
Acceleration	4.5	4.5	4.5
Comfort	4.5	4.5	4.5
Handling	4.5	4	4
Parts	4	3.5b	4
Reliability	3.5w	3.5w	3.5w

b - Targa roofs no longer available from Porsche.   w - Problems linger, see text.

1993 models, manual transmission, continued (2)

	911 Carrera 4 coupe	Carrera 4 Targa	Carrera 4 Cabrio
Acceleration	4.5	4.5	4.5
Comfort	4.5	4.5	4.5
Handling	5	4.5	4
Parts	4	3.5b	4
Reliability	3.5w	3.5w	3.5w

b - Targa roofs no longer available from Porsche.   w - Problems linger, see text.

1993 models, manual transmission, continued (3)

	Turbo S coupe	Turbo 3.6 coupe	911RS America coupe	America Roadster
Acceleration	4.5	5	4	4
Comfort	5	5	2.5	4
Handling	5	5	4.5	4
Parts	4	4	4	4
Reliability	4w	4w	3.5w	3.5w

w - Problems linger, see text.

Ratings continued

1993 models, Tiptronic transmission

	Carrera 2 coupe	Carrera 2 Targa	Carrera 2 Cabrio	America Roadster
Acceleration	3.5	3.5	3.5	3.5
Comfort	4.5	4.5	4.5	4
Handling	4.5	4	4	4
Parts	4	3.5b	4	4
Reliability	3.5w	3.5w	3.5w	3.5w

b - Targa roofs no longer available from Porsche.
w - Problems linger, see text.

1993 Garage Watch

Problems with (and improvements to) Porsche 911 models.

America Roadster introduced.

Turbo 3.6-liter introduced.

Dual distributors connected by belt. If belt breaks engine runs very poorly because one distributor fires correctly and the other fires the nearest plug. Six cylinders fire normally and one sparks all the time. If 964 engine starts running poorly, shut it off as quickly as possible. Problems get more expensive the longer you drive, as from 1989 1/2.

Cylinder head leaks, as from 1989.

Through-bolt O-ring oil leaks, as from 1989.

Dual mass flywheel, as from 1989.

Tiptronic seals, as from 1990.

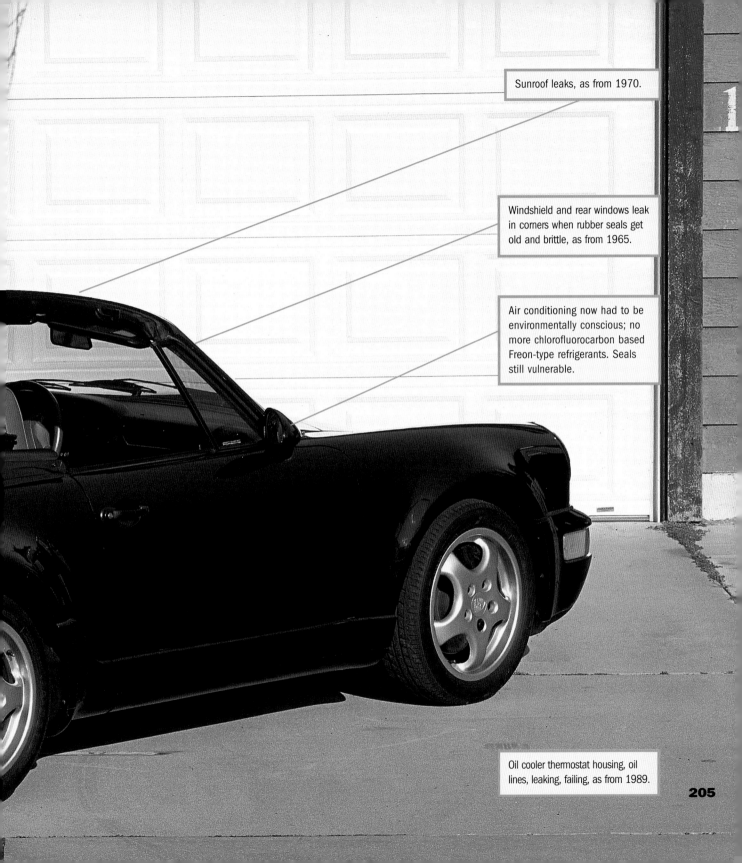

Sunroof leaks, as from 1970.

Windshield and rear windows leak in corners when rubber seals get old and brittle, as from 1965.

Air conditioning now had to be environmentally conscious; no more chlorofluorocarbon based Freon-type refrigerants. Seals still vulnerable.

Oil cooler thermostat housing, oil lines, leaking, failing, as from 1989.

1994

Chapter 34
The New "R" Program—From 964 in December to 993 in January

Porsche trimmed its end-of-model-year 1994 offerings to the Carrera 4 coupe, 3.6-liter Turbo, and limited edition Speedster. Just as Porsche had done with the originals in the mid-1950s and the 1989, this new C2-based model offered only a manually operated cloth top and carried over the twin-hump composite tonneau cover that first appeared on the 1989 model.

Porsche had introduced the all-wheel-drive Carrera 4, the Typ 964, in mid–model year 1989 and the C2 in 1990. Porsche controlled the 219.9 cid engine with the Bosch Digital Motor Electronics (or Motronic) system. It utilized hydraulic timing chain tensioners and ceramic exhaust port liners to reduce maintenance.

Belts often broke on the belt-driven dual distributors, causing one distributor to fire properly and the other to spark each time the plug nearest to where the rotor is stuck.

Perhaps the new model's most notable failure was the dual-mass flywheel, introduced on the 1989 1/2 C4. Created to dampen engine vibration that caused a rattle in the transmission, the flywheel and clutch overheated if the driver slipped the clutch consistently. There are also reports that the clutch disc liner springs lost tension.

The Freudenberg flywheels also incorporated bolted-on starter ring gears that worked loose. Near the end of model year 1992, Porsche replaced the flywheels with the "LUK" update. This most likely has been updated (about $1,100) in any car you look at. To check, let the clutch out slowly and listen for a "clank" that indicates loose or fallen starter gear or secondary flywheel mounting bolts, or a loose or broken heatshield. Do not drive this car.

For the new 964, Porsche retained MacPherson front struts but replaced the rear torsion bars with coil springs and shock absorbers. In an obvious attempt to tame the handling, engineers adapted the 928's rear axle, leading to criticism that they had tamed the cars too much. In fact, with the additional weight up front, all-wheel-drive Carrera 4s understeered and plowed in turns.

In January 1990, Porsche introduced its computer-controlled four-speed Tiptronic transmission that could be operated fully automatically or shifted manually. Some seals leak. Also, in fully automatic operation, the engine does not get worked hard enough.

Responding to environmental concerns, Porsche switched its air conditioning refrigerant from Freon-type CFC coolant to HFC. Some seals on the new system were vulnerable to leakage. In addition, option M573 automatic climate control, introduced in 1989 1/2, brought complicated, troublesome, and expensive problems.

Now that Porsche models are more electronically intensive and mechanically complex, if they are running poorly, you must shut them down as soon as possible. The problems get more expensive the longer you drive them.

By the same token, as a potential buyer, you have fewer worries about the insides of these engines. They don't lose compression at 150,000 miles. You could have a failure if the previous owner ran low on oil, used the wrong plugs, or consistently shifted gears below 3,000 rpm.

The improved reliability of 964, 993, and 996 models has enticed many individuals to buy their first Porsche. But getting first-time owners to run the engine past 4,500 rpm is difficult. The truth is that if you don't run these engines, you are doing more harm than good.

Porsche remedied a number of previous shortcomings with its "new" car, the Typ 993 coupe, introduced in 1994. In March it showed the 993 Cabriolet and in April, U.S. customers began taking delivery of cars Porsche called Carreras.

1994 Specifications "R" Series

Body Designation:		Typ 964 Carrera 4, Carrera 2, Turbo 3.6
Price:		964 Carrera 4 coupe: $73,440
		Cabriolet: $82,215
		964 Carrera 2 coupe: $64,990 (These are carryover 1993 964s badged, licensed, and titled as model year 1994 cars)
		Cabriolet: $74,190
		Speedster: $66,400
		964 Turbo: $95,000
Engine Displacement and Type:		Carrera 4, 2 coupe: Typ M64/01; 3,605 cc, 219.9 cid, SOHC, Bosch Digital Motor Electronic (DME) management system, dual-spark, twin-plugs
		(M64/02, as above, for Tiptronic models.)
		Turbo: Typ M30/50; 3,605 cc, 219.9 cid, SOHC
		Bosch DME, Turbocharger, Intercooler, 3-way catalytic converters
Maximum Horsepower @ rpm:		Carrera 4, 2: 247 SAE @ 6,100 rpm
		Turbo: 355 SAE @ 5,500
Maximum Torque @ rpm:		Carrera 4, 2: 228 ft-lb @ 4,200 rpm
		Turbo: 384 @ 4,200 rpm
Weight:		Carrera 4: 3,340 pounds
		Carrera 2: 3,031 pounds, add 69 pounds for Tiptronic
		Speedster: 3,000 pounds
		Turbo: 3,274 pounds
0–60 mph:		Carrera 4: 5.5 seconds (factory)
		Carrera 2: 5.5 seconds (factory)
		Carrera 2/Tiptronic: 5.5 seconds
		Speedster: 5.5 seconds (factory)
		Turbo: 4.7 seconds (factory)
Maximum Speed:		Carrera 4: 158 mph (factory)
		Carrera Tip: 159 mph (factory)
		Speedster: 162 mph (factory)
		Turbo: 174 mph (factory)
Brakes:		Carrera and Turbo: power venteddisc brakes w/ABS anti-locking brake system
Steering:		force-sensitive hydraulically assisted rack-and-pinion
Suspension:	Front:	MacPherson struts, shock absorbers, lower control arms, coil springs, antisway bar
	Rear:	semi-trailing arms, coil springs with toe correction, antisway bar
Tires:	Front:	Carrera: 205/50ZR16
		Speedster: 205/50ZR17
		Turbo: 225/40ZR18
	Rear:	Carrera: 245/45ZR16
		Speedster: 255/45ZR17
		Turbo: 265/35ZR18
Tire air pressure:		Carrera and Turbo: Front and Rear: 36 psi
		Speedster: Front: 36 psi; Rear: 43 psi
Transmission(s):		Carrera 4: G64/00 5-speed
		Carrera 2: G50/03 5-speed
		Tiptronic: A50/03
		Turbo: G50/52 5-speed w/ZF mechanical limited-slip differential, 20 percent lockup
Wheels:	Front:	Carrera: 7.0Jx16
		Speedster:7.0Jx17
		Turbo: 8.0J-18
	Rear:	Carrera: 9.0Jx16
		Speedster: 9.0Jx17
		Turbo: 10.0J-18

What they said at the time–Porsche for 1994

Speedster, *Road & Track*, September 1993

"Think of the 911 Speedster as a cross between the Carrera 2 Cabriolet and the RS America, the austere, lightweight driver's Porsche, and think of it as the antidote to the 1989 Speedster.

"Obtaining performance superiority was a slam-dunk, duck-soup, piece-of-cake picnic, since the new Speedster is built, note for note, on the still-fresh 911 Carrera 2 configuration, including floorpan, engine, drivetrain, suspension and brakes. About the only carryover from the 1989 Speedster is the windshield, and even that has been more strongly secured within the car's aluminum A-pillars and header."

Parts List for 1994 911s

These are items most commonly replaced during regular maintenance and routine daily operation. Prices quoted are for new factory parts at list price, not including installation labor. NLA means factory parts no longer available, so prices quoted are from aftermarket suppliers.

Engine:

1. Oil filter..................... $15.82
2. Alternator belt set..... $57.14
3. Starter...................... $730.24
4. Alternator................. $942.86
5. Muffler $1,702.05
   (Turbo) $1,659.47
6. Clutch disc............... $284.56

Body:

7. Front bumper............ $1,657.70
8. Left front fender........ $1,780.45
   (Turbo fender
   plus flare) $2,328.77
9. Right rear
   quarter panel............ $2,088.84
10. Front deck lid........... $1,436.00
11. Front deck lid struts.. $18.50 each
12. Rear deck lid struts... $14.24 each
13. Porsche badge,
    front deck lid $38.64
14. Taillight housing
    and lens $186.50
15. Windshield
    with antenna $647.80
16. Windshield weather
    stripping................... $89.69

Interior:

17. Dashboard................ $452.90
18. Shift knob $237.07
19. Interior carpet,
    complete $900.00

Chassis:

20. Front rotor................ $300.76
21. Brake pads,
    front set $208.69
22. Koni rear
    shock absorber.......... $249.24
23. Front wheel $910.61
24. Rear wheel $1,007.92

Ratings

1994 models, manual transmission

	Carrera 2	Cabrio	Speedster	RS America coupe	Carrera 4S coupe
Acceleration		4.5	4.5	4.5	4.5
Comfort	4.5	3	2.5	4.5	
Handling	4.5	4.5	4.5	5	
Parts	4	3.5v	4	4	
Reliability	4w	4w	4w	4w	

v - Speedster "humps" costly to replace.
w - Problems still linger, see text.

Ratings continued

1994 models, manual transmission, continued (2)

	911 Turbo 3.6 coupe
Acceleration	5
Comfort	5
Handling	5
Parts	4
Reliability	4w

w - Problems still linger, see text.

1994 models, Tiptronic transmission

	Carrera 2 Cabriolet	Speedster
Acceleration	4.5	4.5
Comfort	4.5	3
Handling	4.5	4.5
Parts	4	3.5v
Reliability	3.5	3.5

v - Speedster "humps" costly to replace.
w - Problems still linger, see text.

1994 model 993, manual transmission and Tiptronic transmission

	Carrera coupe	Carrera Cabrio
Acceleration	4.5	4.5
Comfort	5	5
Handling	4.5	4
Parts	4	4
Reliability	4	4

209

1994 Garage Watch
Problems with (and improvements to) Porsche 911 models.

Dual distributors on Typ 964 Carrera 2 and 4 models and on 993, connected by belt. If belt breaks engine runs very poorly because one distributor fires correctly and the other fires the nearest plug. Six cylinders fire normally, and one sparks all the time. If 993 engine starts running poorly, shut it off as quickly as possible. Problems get more expensive the longer you drive, as from 1989 1/2.

Still some leakage at sunroof, since 1970.

New exhaust system and new engine management system.

Typ 964 introduced, new engine features maintenance free-hydraulic valve adjusters, six-speed trans, new clutch, flywheel, new front/rear suspensions, revised oil cooler.

Windshields "creak." Revised gluing-in techniques and procedures cured problem.

Leaks continue in windshield and rear window, since 1965.

New climate control with dust/pollen filters.

Ellipsoid headlights replace halogen bulb-type.

Subframe so strong that body shops are advised if plate is bent in crash, they must replace it; it cannot be straightened. Yet . . .
Owners are advised to not fit 18-inch wheels and tires, as these induce body flex, allow wind leaks in side windows.

Not quite as radically new as the 911 had been from the 356 in 1965, Typ 993 Carrera models retained only the roof and the front deck lid of the previous Typ 964 Carrera 2 and 4. Despite the almost countless number of new parts, Porsche cut prices by nearly 10 percent on the Carrera 2 replacement. The Cabrio also dropped in price, and while the rear-wheel-drive Carrera had been introduced in midyear 1994, Porsche supplemented coupe bodies with cabrios for the full 1995 model year, with even larger price cuts.

The list of improvements filled an entire page of Porsche's press release, including new body work; six-speed transmission; new multilink rear suspension with transverse A-arms; revised front suspensions; new wheels; larger front brake discs and pads and Bosch's new ABS 5 anti-lock brake system and cross-drilled front and rear rotors; a new interior ventilation system; quieter exhaust; and a change at the factory to water-based paints for all metallic paints.

Porsche continued the windshield and back glass "floating" mounts, adding a new flush-mounted rubber gasket. Unfortunately, the new, much stiffer chassis hammered the glass, causing some delamination in corners and, in some cases, popping back glass out of the car body. Porsche immediately advised customers to not fit either the optional 17-inch wheel-and-tire package or even larger aftermarket 18-inch wheels and tires, as this further hardened the ride. Midyear it introduced a "Touring" suspension to accommodate the optional 17-inch wheels and tires.

Revisions to the 3.6-liter engine included hydraulic valve lifters; lighter valves, pistons, and connecting rods; reduced flow and acoustically redesigned dual exhausts; and a hot-film airflow sensor to better meter fuel mix. Weissach was also able to coax another 23 horsepower out of the engine.

To protect their owners' investments, Porsche also introduced the "Electronic Immobilizer," called "Drive-Block," that acts on the Bosch engine management DME system.

As an option, Porsche introduced a new Automatic Brake Differential (ABD) traction control system that automatically applies braking to the one rear wheel that is spinning during acceleration. This allows the driver to apply maximum engine torque on marginal road conditions.

Where Typ 964 Carrera 4 models had relied on constant velocity joints and a complicated computer system to transmit power to the front tires, Porsche simplified the mechanism greatly by using a viscous center clutch, eliminating the sluggish front drive sensations.

The Weissach rear suspension developed for the 928 and adapted to the 964 Carrera 2 and 4 led to improvements in the rear suspension on the Typ 933. A subframe carries both lower and upper A-arms with additional transverse links. This rear subframe and the front subframe are so strong that if these plates are bent in a crash, they cannot be straightened. They must be replaced.

So drivers need never lift their hands from the steering wheel, Porsche was the first automaker to place the shifter mechanisms on the steering wheel hubs, connected electronically to the four-speed gearbox. Porsche retained the center tunnel-mounted shifter as well. The Tiptronic and Tiptronic S were available on two-wheel-drive cars only, but the S steering wheel controls could be refitted to earlier 1995 Tiptronic-equipped cars.

Accommodating a perpetual desire for turbocharged cars, Porsche announced a new version, equipped with twin turbochargers and all-wheel drive. In short, Porsche had reproduced—and updated—its benchmark stretching 1986 model 959. And for those truly power hungry, Porsche fitted the car with a 150-watt 10-speaker Becker AM-FM stereo cassette or CD audio system. As impressive as all that was, it was its braking that stunned the imagination.

In a final note, at the end of model year 1995, Porsche discontinued manufacture of both water-cooled front engine models, the four-cylinder Typ 968, and the luxurious Typ 928 GTS. Prices of both cars had risen far beyond their perceived value, and sales had dropped to nearly negligible numbers. With their disappearance, Ernst Fuhrmann's plans for Porsche In The Future slipped quietly away.

1995 Specifications "S" Series

Body Designation:		Typ 993 Carrera, Carrera 4
Price:		911 Carrera 4 coupe: $65,900
		Cabriolet: $74,200
		911 Carrera 2 coupe: $59,900
		Cabriolet: $68,200, add for Tiptronic: $3,265
Engine Displacement and Type:		911 Carrera 4, 2 coupe: Typ M64/07; 3,605 cc, 219.9 cid, SOHC, Bosch Digital Motor Electronic (DME) management system, dual-spark, twin-plugs (M64/08, as above, for Tiptronic models.)
Maximum Horsepower @ rpm:		Carrera, 4: 270 SAE @ 6,100 rpm
Maximum Torque @ rpm:		Carrera, 4: 243 ft-lb @ 5,000 rpm
Weight:		Carrera (coupe and cabriolet): 3,064 pounds (factory)
		Carrera 4: 3,175 pounds, add 56 pounds for Tiptronic
0–60 mph:		Carrera: 5.4 seconds (factory)
		Carrera 4: 5.4 seconds (factory)
		Carrera Tiptronic: 6.4 seconds (factory)
Maximum Speed:		Carrera: 168 mph (factory)
		Carrera Tip: 165 mph (factory)
Brakes:		Carrera: power, vented, cross-drilled, dual circuit, 4-piston aluminum alloy fixed calipers, ABS 5 antilocking brake system
Steering:		force-sensitive hydraulically assisted rack-and-pinion
Suspension:	Front:	MacPherson struts, lower control arms, coil springs, antisway bar
	Rear:	multilink, self-stabilizing toe characteristics, antisway bar
Tires:	Front:	Carrera: 205/55ZR16
		optional: 205/50ZR17
	Rear:	Carrera: 245/45ZR16
		optional: 255/40ZR17
Tire air pressure:		All models: front and rear: 36 psi
Transmission(s):		Carrera 2, 4: G50/20 6-speed
		Tiptronic: A50/05
Wheels:	Front:	Carrera: 7.0Jx17
		Turbo: 8.0Jx18
	Rear:	Carrera: 9.0Jx17
		Turbo: 10.0Jx18

What they said at the time–Porsche for 1995

911 Carrera, *Road & Track*, January 1994

"Over and over again Stuttgart has introduced newer cars and tended to downplay the expensive-to-build 911. But, ultimately, it was the front engine cars that foundered, while the 911 blithely sailed on.

"Is there a message here? In slow economic times, don't argue with the customers. . .

"To me, the image of Porsche engineers fussing over their anachronistic 3.6-liter air-cooled flat six has a wonderful air of mystery about it, like some secretive Druid ceremony."

Parts List for 1995 911s (993)

These are items most commonly replaced during regular maintenance and routine daily operation. Prices quoted are for new factory parts at list price, not including installation labor. NLA means factory parts no longer available, so prices quoted are from aftermarket suppliers.

Engine:

1. Oil filter set.............. $54.30
2. Alternator belt set..... $57.14
3. Starter...................... $1,287.90
4. Alternator................. $1035.80
5. Muffler..................... $712.32 each
6. Clutch disc............... $284.56

Body:

7. Front bumper............. $1,666.11
8. Left front fender........ $2,641.01
9. Right rear quarter panel............ $3,309.82
10. Front deck lid........... $1,797.09
11. Front deck lid struts.. $37.23 each
12. Rear deck lid struts... $32.88 each
13. Porsche badge, front deck lid............ $142.29
14. Taillight housing and lens................... $258.18
15. Windshield with antenna............. $1,418.04
16. Windshield weather stripping................... $109.50 set

Interior:

17. Dashboard................. $2,888.65
18. Shift knob $217.05
19. Interior carpet, complete $900.00

Chassis:

20. Front rotor................ $506.50
21. Brake pads, front set.................... $264.48
22. Rear shock absorber... $397.67
23. Front wheel $707.37
24. Rear wheel $1,012.22

Ratings

1995 model 993, manual transmission

	Carrera coupe	Carrera Cabrio	Turbo coupe
Acceleration	4.5	4.5	5
Comfort	4.5	4.5	5
Handling	4.5x	4x	5
Parts	4.5	4.5	4.5
Reliability	4.5	4.5	4.5

x - 17-inch wheels cause problems, see text.

1995 model 993, Tiptronic S transmission

	Carrera coupe	Carrera Cabrio
Acceleration	4	4
Comfort	4.5	4.5
Handling	4.5x	4x
Parts	4.5	4.5
Reliability	4.5	4.5

x - 17-inch wheels cause problems, see text.

1995 Garage Watch
Problems with (and improvements to) Porsche 911 models.

Porsche introduced a new model heralding countless changes. U.S. rear-wheel-drive coupe and cabriolet versions arrived in April 1994 as an early-release 1995 model year car. Virtually everything on the car was new except for the front deck lid and the roof. During the full 1995 model year, Porsche brought along its all-new C4 models. The new car used a viscous couple and central shaft to transmit power to the front end. A much lighter system, it weighed about 110 pounds instead of the 220 that the 964 system had weighed. The Tiptronic S automatic transmission debuted with fingertip gear change rocker buttons on the steering wheel. A new Electronic Immobilizer alarmed the car and disabled the engine through the Bosch DME system.

Steering rack leaks, boots tear.

For cabriolets, check to be certain Cabriolet Top Service Action has been done.

Monroe shocks too soft, leak.

Windshield delaminates at antenna entry.

Front control arm bushings crack. Creates steering shimmy.

On-Board Diagnostics Generation One provides one oxygen sensor and two catalytic converters. Valve guide leaks can foul sensor, possibly killing it and catalytic converters.

Gearbox selector shaft seal leaks. Slave cylinder creaks, hose leaks.

Watch aging Bosch Digital Motor Electronics module relay.

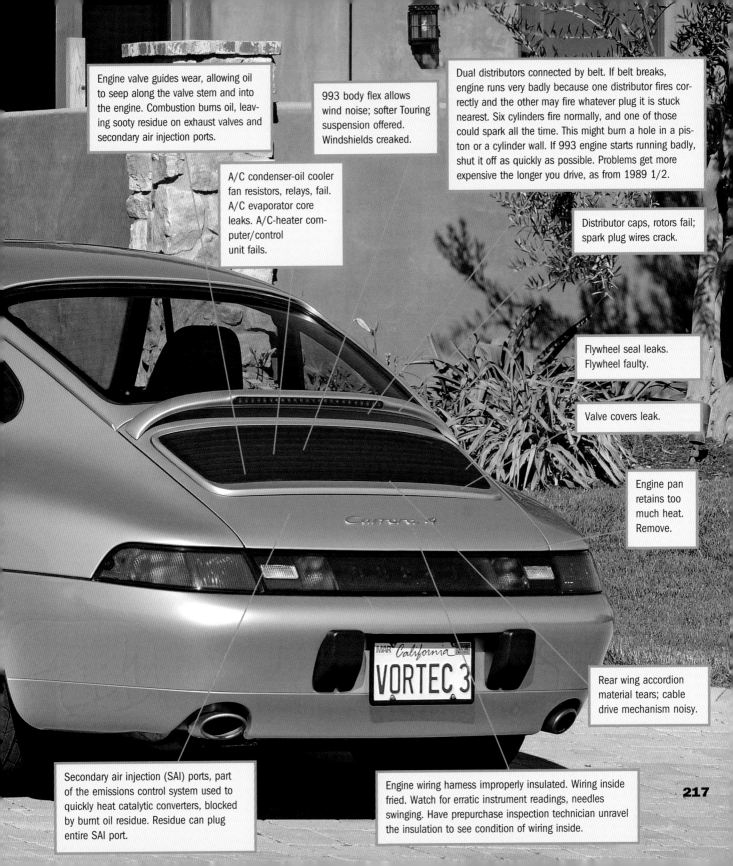

Engine valve guides wear, allowing oil to seep along the valve stem and into the engine. Combustion burns oil, leaving sooty residue on exhaust valves and secondary air injection ports.

993 body flex allows wind noise; softer Touring suspension offered. Windshields creaked.

A/C condenser-oil cooler fan resistors, relays, fail. A/C evaporator core leaks. A/C-heater computer/control unit fails.

Dual distributors connected by belt. If belt breaks, engine runs very badly because one distributor fires correctly and the other may fire whatever plug it is stuck nearest. Six cylinders fire normally, and one of those could spark all the time. This might burn a hole in a piston or a cylinder wall. If 993 engine starts running badly, shut it off as quickly as possible. Problems get more expensive the longer you drive, as from 1989 1/2.

Distributor caps, rotors fail; spark plug wires crack.

Flywheel seal leaks. Flywheel faulty.

Valve covers leak.

Engine pan retains too much heat. Remove.

Rear wing accordion material tears; cable drive mechanism noisy.

Secondary air injection (SAI) ports, part of the emissions control system used to quickly heat catalytic converters, blocked by burnt oil residue. Residue can plug entire SAI port.

Engine wiring harness improperly insulated. Wiring inside fried. Watch for erratic instrument readings, needles swinging. Have prepurchase inspection technician unravel the insulation to see condition of wiring inside.

1996

Chapter 36
The "New" T Program—
Turbos, Targas,
and C4S! Whew!

The Typ 993 Carrera, Carrera 4, and Turbo models retained only the roof and the front deck lid from previous Typ 964 Carrera 2 and 4 cars. Improvements included new body work utilizing poly-ellipsoid low beam and variable-focus high beam headlights; a six-speed transmission; new multilink rear suspension with transverse A-arms; revised front suspensions; new wheels; larger front brake discs and pads, along with Bosch's new ABS 5 anti-lock brake system and cross-drilled front and rear rotors; a new interior ventilation system, with dust and pollen filters; quieter exhaust; and a change at the factory to water-based paints for all metallic paints.

For model year 1996, to resolve the penchant for its large glass surfaces to creak and leak, Porsche began to glue in the windshield and back glass, just as it had done with all 911s until the Typ 964 C4 was introduced in midyear 1989. This change back to glue allowed Porsche to introduce optional 18-inch wheels and tires on all two-wheel-drive Carrera models.

Substantial revisions to the 3.6-liter engine included using hydraulic valve lifters; lighter valves, pistons, and connecting rods; reduced flow (yet acoustically redesigned) dual exhausts; and a hot-film airflow sensor to better meter the fuel mix for increased engine output at higher fuel efficiency. For model year 1996, Weissach introduced its Varioram induction system, which raised output from 270 horsepower SAE net to 282 SAE net at 6,300, and increasing torque from 243 to 250 ft-lb SAE net at 5,250 rpm.

To protect owners' investments, Porsche also introduced the "Electronic Immobilizer," called "Drive-Block," that acts on the Bosch engine management DME system.

To protect owners' exuberance, Porsche introduced a new Automatic Brake Differential (ABD) traction control system as an option. This automatically applies braking to the one rear wheel that is spinning during acceleration up to 44 miles per hour. This allows the driver to apply maximum engine torque to marginal road conditions. On six-speed manual transmission cars, the ABD is included with the optional limited-slip transmission, incorporating a 25 percent differential lockup on acceleration and 40 percent on deceleration. It was standard on the Carrera 4 models.

Where Typ 964 Carrera 4 models had relied on constant velocity joints and a complicated computer system to transmit power to the front tires, Porsche simplified the mechanism greatly by using a viscous center clutch, eliminating the sluggish front drive sensations. Porsche made good use of this improvement with its new turbo-bodied C4S model.

The Weissach rear suspension developed for the 928 models and adapted to the 964 Carrera 2 and 4 automobiles led to improvements in the rear suspension on this new Typ 933. A subframe carries both lower and upper A-arms with additional transverse links. This rear subframe and the front subframe are so strong that if these plates are bent in a crash, they cannot be straightened. They must be replaced.

The Targa returned to Porsche's line-up in 1996, but it was not without problems. Porsche innovated an all-glass roof that slid down inside the rear window. As these 1996 models reached two years of age and more, tops began seizing, jamming open or closed or somewhere in between. The repair is very complicated and time consuming, and therefore very expensive.

Porsche returned the Turbo to its line-up as a 1996 model late in model year 1995. The new version incorporated twin turbochargers and dual intercoolers and all-wheel drive. In short, Porsche had reproduced—and updated—its 1986 stretching model 959, for mere mortals. Where the 959s, produced from 1986 through 1988 with all-wheel drive and 450 horsepower had sold in the neighborhood of $240,000 at the factory (and were not, until very recently, even legal in the United States) this new car was fully U.S. legal, provided 400 horsepower, and sold for $105,000. Porsche said it was capable of a top speed of 180 miles per hour. And for those truly power hungry, Porsche fitted the car with a 150-watt 10-speaker Becker AM-FM stereo cassette-or-CD audio system. As impressive as all that was, it was its braking that stunned the imagination. While 0 to 60 miles per hour took 4.4 seconds, with its huge rotors and Bosch's ABS 5 anti-lock system, 60 miles per hour to standstill took 2.6 seconds.

1996 Specifications "T" Series

Body Designation:		Typ 993 Carrera coupe, Carrera Targa, Carrera Cabriolet, Carrera 4S coupe, Carrera 4 Cabriolet, Turbo 3.6
Price:		911 Carrera coupe: $63,750
		Targa: $70,750
		Cabriolet: $73,000
		911 Carrera 4S coupe: $73,000
		Carrera 4 Cabriolet: $78,350, add for Tiptronic: $ 3,265
		993 Turbo: $105,000
Engine Displacement and Type:		911 Carrera coupe, 4S: Typ M64/23; 3,608 cc, 219.9 cid, SOHC, Bosch Digital Motor Electronic (DME) management system, dual-spark, twin-plugs (M64/24, as above, for Tiptronic models.)
		Turbo: Typ M64/60; 3,608 cc, 219.9 cid, SOHC
		Bosch M5.2, port fuel injection, turbocharger, Intercooler, 3-way catalytic converters
Maximum Horsepower @ rpm:		Carrera, 4S: 282 SAE @ 6,300 rpm
		Turbo: 400 SAE @ 5,750 rpm
Maximum Torque @ rpm:		Carrera 4, 2: 243 ft-lb @ 5,000 rpm
		Turbo: 400 @ 4,500
Weight:		Carrera coupe: 3,064 pounds
		Carrera 4S coupe: 3,065 pounds
		Targa: 3,130 pounds, add 56 pounds for Tiptronic
		Turbo: 3,307 pounds
0–60 mph:		Carrera: 5.3 seconds (factory)
		Carrera 4S: 5.2 seconds (factory)
		Carrera Tiptronic: 6.3 seconds (factory)
		Turbo: 4.4 seconds (factory)
Maximum Speed:		Carrera: 171 mph (factory)
		Carrera 4S: 168 mph (factory)
		Tiptronic: 168 mph (factory)
		Turbo: 180 mph (factory)
Brakes:		Carrera and Turbo: power, vented, cross-drilled, dual-circuit, four- piston aluminum alloy fixed caliper, ABS 5 anti-locking brake system
Steering:		force-sensitive hydraulically assisted rack-and-pinion
Suspension:	Front:	MacPherson struts, telescoping shock absorbers, lower control arms, coil springs, antisway bar
	Rear:	multilinks, self-stabilizing toe characteristics, antisway bar
Tires:	Front:	Carrera and Targa: 205/50ZR17
		optional Carrera: 225/40ZR18
		Turbo: 225/40ZR18
	Rear:	Carrera and Targa: 255/40ZR17
		optional Carrera: 265/35ZR18
		Turbo: 285/30ZR18
Tire air pressure:		All models, front and rear: 36 psi
Transmission(s):		Carrera, 4S: G50/20 6-speed
		Tiptronic: A50/05
		Turbo: G64/51 6-speed four-wheel drive, w/ZF mechanical limited-slip differential, 20 percent lockup; Differential (ABD) traction system
Wheels:	Front:	Carrera: 7.0Jx17
		optional: 8.0Jx18
		Turbo: 8.0Jx18
	Rear:	Carrera: 9.0Jx17
		optional: 10.0Jx18
		Turbo: 10.0Jx18

What they said at the time–Porsche for 1996

1996 Carrera Targa, *Complete Car***, November 1995**

"The thing that strikes you first when you drive the Targa is how uncompromised it feels. Its reinforced Cabrio-based body shell gives the impression of being every bit as still as the coupe's. In fact it's only 75 percent of the way there and some 65 pounds heavier, but it doesn't flex or shimmy over bumps. Porsche's chassis engineers have softened up the suspension to precisely the right degree that not only does it not draw attention to the slight shortfall in rigidity, it also gives the Targa arguably the most comfortable ride in the range."

Parts List for 1996 911s

These are items most commonly replaced during regular maintenance and routine daily operation. Prices quoted are for new factory parts at list price, not including installation labor. NLA means factory parts no longer available, so prices quoted are from aftermarket suppliers.

Engine:

1. Oil filter set.............. $54.30
2. Alternator belt set..... $57.14
3. Starter...................... $1,287.90
4. Alternator................. $1,035.80
5. Muffler .... $712.32 each
    (Turbo) .... $1,049.55 each
6. Clutch disc............... $284.56

Body:

7. Front bumper............. $1,666.11
8. Left front fender........ $2,641.01
    (Turbo fender
    plus flare) $3,226.70
9. Right rear
    quarter panel............. $3,309.82
10. Front deck lid $1,797.09
11. Front deck lid struts.. $37.23 each
12. Rear deck lid struts... $32.88 each
13. Porsche badge,
    front deck lid............. $142.29
14. Taillight housing
    and lens $258.18
15. Windshield
    with antenna $1,418.04
16. Windshield weather
    stripping................... $109.50 set

Interior:

17. Dashboard................. $2,888.65
18. Shift knob $217.05
19. Interior carpet,
    complete $900.00

Chassis:

20. Front rotor set $506.50
21. Brake pads,
    front set $264.48
22. Rear shock absorber... $397.67
23. Front wheel $707.37
24. Rear wheel $1,012.22

Ratings

1996 models, manual transmission

	Carrera coupe	Carrera Targa	Carrera Cabrio
Acceleration	4.5	4.5	4.5
Comfort	4.5	4.5	4.5
Handling	4.5	4	4
Parts	4.5	4.5	4.5
Reliability	4.5	2y	4.5

y - Targa glass roof jams, see text.

1996 models, Tiptronic S transmission

	Carrera coupe	Carrera Targa	Carrera Cabrio
Acceleration	4	4	4
Comfort	4.5	4.5	4.5
Handling	4.5	3.5	4
Parts	4	4	4
Reliability	4.5	2y	4.5

y - Targa glass roof jams, see text.

1996 models, manual transmission

	Turbo coupe	Carrera 4S coupe
Acceleration	5	4.5
Comfort	5	4.5
Handling	5	5
Parts	5	5
Reliability	5	5

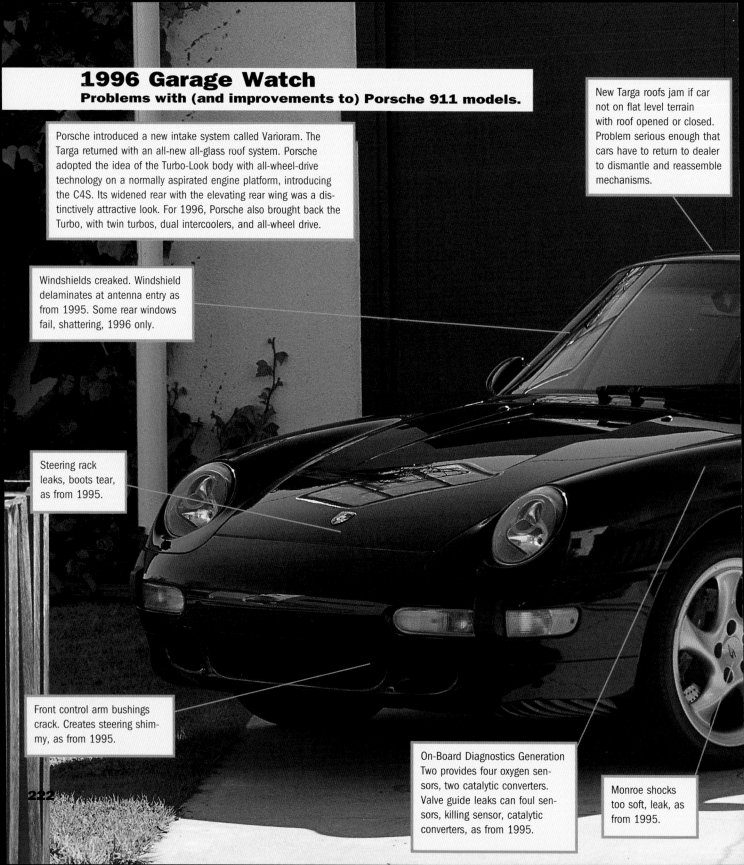

1996 Garage Watch
Problems with (and improvements to) Porsche 911 models.

Porsche introduced a new intake system called Varioram. The Targa returned with an all-new all-glass roof system. Porsche adopted the idea of the Turbo-Look body with all-wheel-drive technology on a normally aspirated engine platform, introducing the C4S. Its widened rear with the elevating rear wing was a distinctively attractive look. For 1996, Porsche also brought back the Turbo, with twin turbos, dual intercoolers, and all-wheel drive.

New Targa roofs jam if car not on flat level terrain with roof opened or closed. Problem serious enough that cars have to return to dealer to dismantle and reassemble mechanisms.

Windshields creaked. Windshield delaminates at antenna entry as from 1995. Some rear windows fail, shattering, 1996 only.

Steering rack leaks, boots tear, as from 1995.

Front control arm bushings crack. Creates steering shimmy, as from 1995.

On-Board Diagnostics Generation Two provides four oxygen sensors, two catalytic converters. Valve guide leaks can foul sensors, killing sensor, catalytic converters, as from 1995.

Monroe shocks too soft, leak, as from 1995.

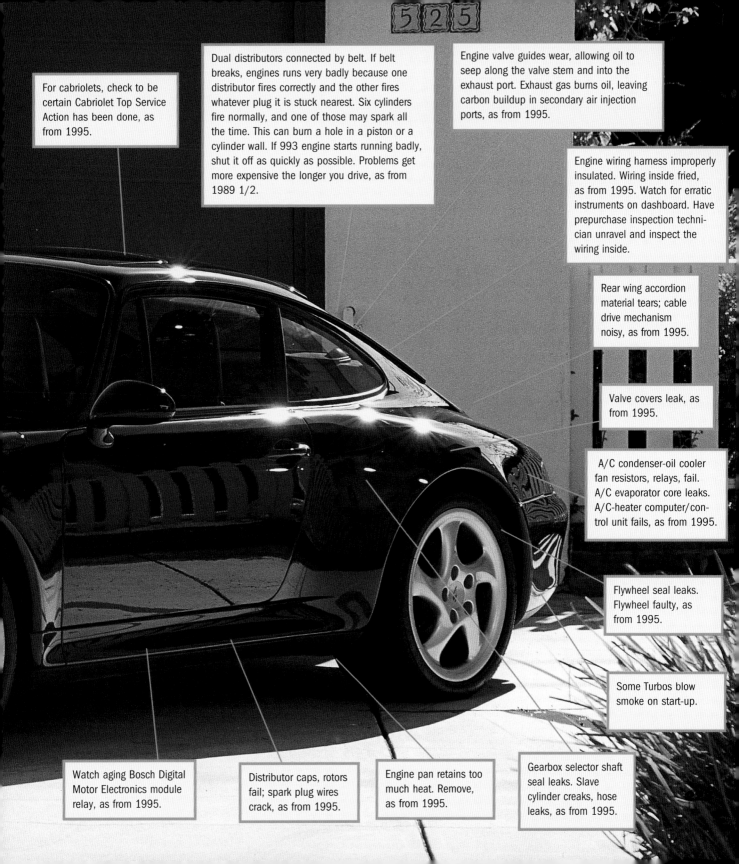

5 2 5

For cabriolets, check to be certain Cabriolet Top Service Action has been done, as from 1995.

Dual distributors connected by belt. If belt breaks, engines runs very badly because one distributor fires correctly and the other fires whatever plug it is stuck nearest. Six cylinders fire normally, and one of those may spark all the time. This can burn a hole in a piston or a cylinder wall. If 993 engine starts running badly, shut it off as quickly as possible. Problems get more expensive the longer you drive, as from 1989 1/2.

Engine valve guides wear, allowing oil to seep along the valve stem and into the exhaust port. Exhaust gas burns oil, leaving carbon buildup in secondary air injection ports, as from 1995.

Engine wiring harness improperly insulated. Wiring inside fried, as from 1995. Watch for erratic instruments on dashboard. Have prepurchase inspection technician unravel and inspect the wiring inside.

Rear wing accordion material tears; cable drive mechanism noisy, as from 1995.

Valve covers leak, as from 1995.

A/C condenser-oil cooler fan resistors, relays, fail. A/C evaporator core leaks. A/C-heater computer/control unit fails, as from 1995.

Flywheel seal leaks. Flywheel faulty, as from 1995.

Some Turbos blow smoke on start-up.

Watch aging Bosch Digital Motor Electronics module relay, as from 1995.

Distributor caps, rotors fail; spark plug wires crack, as from 1995.

Engine pan retains too much heat. Remove, as from 1995.

Gearbox selector shaft seal leaks. Slave cylinder creaks, hose leaks, as from 1995.

The Typ 993 Carrera, Carrera 4, and Turbo models retained only the roof and the front deck lid from previous Typ 964 Carrera 2 and 4 cars. The list of improvements included new bodywork utilizing poly-ellipsoid low beam and variable-focus high beam headlights; six-speed gearbox; new multilink rear suspension with transverse A- arms; revised front suspensions; new wheels; larger front brake discs and pads, along with Bosch's new ABS 5 anti-lock brake system, and cross-drilled front and rear rotors; a new interior ventilation system, with dust and pollen filters; quieter exhaust; and a change at the factory to water-based paints for all metallic paints.

Porsche had placed the C2 and C4 windshields and back glass in "floating" mounts and continued this with the new Carrera, adding a new flush-mounted rubber gasket. Unfortunately, these large glass surfaces continued to creak and leak. For model year 1996, to resolve the problem, it began to glue in the windshield and back glass as it had done until the C4 was introduced in mid–year 1989. This allowed Porsche to introduce optional 18-inch wheels and tires on all two-wheel-drive Carrera models.

Substantial revisions to the 3.6-liter engine included using hydraulic valve lifters; lighter valves, pistons, and connecting rods; reduced flow (yet acoustically redesigned) dual exhausts; and a hot-film airflow sensor to better meter fuel mix for increased engine output at higher fuel efficiency. For model year 1996, Weissach introduced its Varioram induction system, which eked another 12 horsepower out of the 3.6-liter engine, raising output from 270 horsepower SAE net to 282 SAE net at 6,300, and increasing torque from 243 to 250 ft-lb SAE net at 5,250 rpm.

To protect owners' exuberance, Porsche introduced a new Automatic Brake Differential (ABD) traction control system as an option. This automatically applies braking to the one rear wheel that is spinning during acceleration up to 44 miles per hour. This allows the driver to apply maximum engine torque to marginal road conditions. On six-speed manual transmission cars, the ABD is included with the optional limited-slip transmission, incorporating a 25 percent differential lockup on accelerating and 40 percent on deceleration. It was standard on the Carrera 4 models.

Where Typ 964 Carrera 4 models had relied on constant velocity joints and a complicated computer system to transmit power to the front tires, Porsche simplified the mechanism greatly by using a viscous center clutch, eliminating the sluggish front drive sensations. Porsche made good use of this improvement with its new turbo-bodied C4S coupe. (Porsche added a wide-bodied C2S coupe for mid–model year 1997, introduced in March.)

The Weissach rear suspension developed for the 928 models and adapted to the 964 Carrera 2 and 4 automobiles led to improvements in the rear suspension on this new Typ 933. A subframe carries both lower and upper A-arms with additional transverse links. This rear subframe and the front subframe are so strong that if these plates are bent in a crash, they cannot be straightened. They must be replaced.

In mid-1997, Porsche added an even more tantalizing package to its Turbo line-up, the Turbo S, introduced at the New York Auto Show in March 1997. The new version carried over the standard model's twin turbochargers and dual intercoolers and all-wheel drive. With the Turbo, Porsche had reproduced—and updated—its stretching 1986 model 959, for mere mortals. Where the 1986 through 1988 959s with all-wheel drive and 450 horsepower had sold in the neighborhood of $240,000 at the factory (and were not, until very recently, even legal in the United States) this new car was fully U.S. legal, provided 400 horsepower and sold for $105,000. Porsche said it was capable of a top speed of 180 miles per hour. The Turbo S sold for $150,000, and produced 424 conservatively rated horsepower. Porsche fitted both Turbos with a 150-watt 10-speaker Becker AM-FM stereo cassette-or-CD audio system.

In a hint of things to come with the 911 models, Porsche introduced its newest water-cooled model, the mid-engined Boxster, its first new sports car in 19 years. Using an exclusively Porsche 2.5-liter six-cylinder engine, Porsche's Boxster developed 201 horsepower. Porsche had determined that, in at least one way, Ernst Fuhrmann was correct: Water-cooled engines were easier to develop while maintaining ever stricter environmental and noise pollution controls.

1997 Specifications "V" Series

Body Designation:		Typ 993 Carrera coupe, Carrera Targa, Carrera and Carrera 4 Cabriolet, Carrera 4S coupe, Turbo, Turbo S
Price:		911 Carrera coupe: $63,750
		Cabriolet: $73,000
		Targa: $70,750
		911 Carrera 4S coupe: $73,000
		Cabriolet: $78,350, add for Tiptronic: $ 3,265
		Turbo: $105,000
		Turbo S: $150,000
Engine Displacement and Type:		911 Carrera 4, 2 coupe: Typ M64/23; 3,608 cc, 219.9 cid, SOHC, Bosch Digital
		Motor Electronic (DME) management system, dual-spark, twin-plugs (M64/24,
		as above, for Tiptronic models)
		Turbo: Typ M64/60; 3,608 cc, 219.9 cid, SOHC
		Bosch M5.2, port fuel injection, twin turbochargers, intercooler, 3-way catalytic converters
Maximum Horsepower @ rpm:		Carrera coupe, S, 4S:
		282 SAE @ 6,300 rpm
		Turbo: 400 SAE @ 5,700
		Turbo S: 424 SAE @ 5,750
Maximum Torque @ rpm:		Carrera coupe, S, 4S:
		250 ft-lb @ 5,250 rpm
		Turbo: 400 @ 4,500
		Turbo S: 400 @ 4,500
Weight:		Carrera coupe: 3,064 pounds
		Carrera 4S: 3,197 pounds
		Targa: 3,130 pounds, add 56 pounds for Tiptronic
		Turbo, S: 3,307 pounds (factory)
0–60 mph:		Carrera, S: 5.3 seconds (factory)
		Carrera 4S: 5.2 seconds (factory)
		Carrera Tiptronic: 6.3 seconds (factory)
		Turbo: 4.4 seconds (factory)
		Turbo S: 4.1 seconds (factory)
Maximum Speed:		Carrera: 171 mph (factory)
		Carrera 4S: 168 mph
		Carrera Tip: 168 mph (factory)
		Turbo: 180 mph (factory)
		Turbo S: 189 mph (factory)
Brakes:		Carrera: power, vented, cross-drilled, four-piston aluminum alloy fixed caliper,
		ABS 5 anti-locking brake system
Steering:		hydraulically assisted rack-and-pinion
Suspension:	**Front:**	MacPherson struts with telescoping shock absorbers, coil springs, lower control arms, antisway bar
	Rear:	Multilink, self-stabilizing toe characteristics, antisway bar
Tires:	**Front:**	Carrera and Targa: 205/55ZR17
		Carrera optional: 225/40ZR18
		Turbo: 225/40ZR18
	Rear:	Carrera and Targa: 245/45ZR17
		optional: 265/35ZR18
		Turbo: 285/30ZR18
Tire air pressure:		All models, front and rear: 36 psi
Transmission(s):		Carrera: G50/20 6-speed
		Tiptronic: A50/05
		Turbo: G64/51 6-speed four-wheel drive, w/ZF mechanical limited-slip differential,
		20 percent lockup; Automatic Brake Differential (ABD) traction system

1997 Specifications "V" Series

Wheels:	Front:	Carrera: 7.0Jx17
		optional: 8.0Jx18
		Turbo: 8.0Jx18
	Rear:	Carrera: 9.0Jx17
		optional: 10.0Jx18
		Turbo: 10.0Jx18

What they said at the time–Porsche for 1997

911 Carrera S, *Motor Sport*, March 1997

"A standard 911, when driven hard across undulating terrain, can feel a little soft and lacking in ultimate body control through tough dips and humps. In the Carrera S, as long as it's dry, this feeling of slight unease has been eradicated entirely and replaced by a sense of implacable commitment to the job ahead. It attacks corners with added verve and clings to the apex with a determination remarkable even by the impressive standards of the breed."

Parts List for 1997 911s

These are items most commonly replaced during regular maintenance and routine daily operation. Prices quoted are for new factory parts at list price, not including installation labor. NLA means factory parts no longer available, so prices quoted are from aftermarket suppliers.

Engine:

1. Oil filter set............... $54.30
2. Alternator belt set $57.14
3. Starter...................... $1,287.90
4. Alternator................. $1,035.80
5. Muffler .... $712.32 each
   (Turbo) .... $1,049.55 each
6. Clutch disc $284.56

Body:

7. Front bumper............ $1,666.11
8. Left front fender........ $2,641.01
   (Turbo fender
   plus flare) $3,226.70
9. Right rear
   quarter panel............ $3,309.82
10. Front deck lid........... $1,797.09
11. Front deck lid struts .. $37.23 each
12. Rear deck lid struts... $32.88 each
13. Porsche badge,
    front deck lid............ $142.29
14. Taillight housing
    and lens $258.18
15. Windshield
    with antenna $1,418.04
16. Windshield weather
    set stripping............. $109.50

Interior:

17. Dashboard................ $2,888.65
18. Shift knob $217.05
19. Interior carpet,
    complete $900.00

Chassis:

20. Front rotor set.......... $506.50
21. Brake pads,
    front set $264.48
22. Rear shock absorber... $397.67
23. Front wheel $707.37
24. Rear wheel $1,012.22

Ratings

1997 models, manual transmission

	Carrera S coupe	Carrera Targa	Carrera 4 Cabrio	Carrera 4S
Acceleration	4.5	4.5	4.5	4.5
Comfort	4.5	4.5	4.5	4.5
Handling	4.5	3.5	3.5	5
Parts	4	4	4	4
Reliability	4.5	2y	4.5	4.5

y - Targa glass roofs jam, see text.

Ratings continued

1997 models, manual transmission, continued (2)

	911 Turbo	911 Turbo S
Acceleration	5	5
Comfort	5	5
Handling	5	5
Parts	4	4
Reliability	5	5

1997 models, Tiptronic S transmission

	Carrera S coupe	Carrera Targa
Acceleration	4	4
Comfort	4.5	4.5
Handling	4.5	3.5
Parts	4	4
Reliability	4.5	2y

y - Targa glass roofs jam, see text.

1997 Garage Watch
Problems with (and improvements to) Porsche 911 models.

Tiptronic S transmission with shifter on steering wheel; still suffers seals liability, since 1990.

For cabriolets, check to be certain Cabriolet Top Service Action has been done, as from 1995.

Targa discontinued, then returns late in model year with entirely new system.

Dual distributors connected by belt. If belt breaks, engines runs very badly because one distributor fires correctly and the other fires whatever plug it is stuck nearest. This can cause serious and expensive problems, as from 1989 1/2 and 1995.

Rear wing accordion material tears; cable drive mechanism noisy, as from 1995.

Valve covers leak, as from 1995.

Engine valve guides wear, allowing oil to seep along the valve stem and into the exhaust port. Exhaust gas burns oil, leaving carbon buildup in secondary air injection ports. By the time Check Engine light appears, problem may be serious, as from 1995.

Some Turbos blow smoke on start-up, as from 1996.

Engine pan retains too much heat. Remove, as from 1995.

Distributor caps, rotors fail; spark plug wires crack, as from 1995.

Windshield and rear window outer seals creaked. Windshield delaminates at antenna entry, as from 1995.

Few changes. Internally Porsche was putting final touches on next generation 911 with water cooling and another new body. Weissach engineers developed a fix for fried wiring harnesses above the engine by using different materials from same manufacturer. Otherwise, a series of problems outlined here were looming on the horizon when cars hit 20,000, 30,000, and 40,000 miles.

Gearbox selector shaft seal leaks. Slave cylinder creaks, hose leaks. as from 1995.

Steering rack leaks, boots tear, as from 1995.

Front control arm bushing crack. Creates steering shimmy, as from 1995.

Monroe shocks too soft, leak, as from 1995.

Watch aging Bosch Digital Motor Electronics module relay, as from 1995.

On-Board Diagnostics Generation Two provides four oxygen sensors, two catalytic converters. Valve guide leaks can foul sensors, killing sensor, catalytic converters, as from 1995.

A/C condenser-oil cooler fan resistors, relays, fail. A/C evaporator core leaks. A/C-heater computer/control unit fails, as from 1995.

Flywheel seal leaks. Flywheel faulty, as from 1995.

1998

Chapter 38
The "New" W Program—
The Future Is Coming,
Next Year!

Porsche introduced the new water-cooled Typ 996 in Europe as a 1998 model and showed it in the United States for sale as a 1999 model year offering. For U.S. customers, the Typ 993 Carrera S coupe, C4S coupe, Carrera, C4 Cabriolet, and the glass-roofed Targa remained in production. The company dropped the Turbo and Turbo S models. The 993 coupes had retained only the roof and the front deck lid from previous Typ 964 Carrera 2 and 4 cars. The list of improvements filled an entire page of Porsche's press release, including new body work utilizing poly-ellipsoid low beam and variable-focus high beam headlights; six-speed gearbox; new multilink rear suspension with transverse A-arms; revised front suspensions; new wheels; larger front brake discs and pads, along with Bosch's new ABS 5 anti-lock brake system and cross-drilled front and rear rotors; a new interior ventilation system, with dust and pollen filters; quieter exhaust; and a change at the factory to water-based paints for all metallic paints.

Porsche substantially revised the 3.6-liter engine, now using hydraulic valve lifters; lighter valves, pistons, and connecting rods; reduced flow (yet acoustically redesigned) dual exhausts; and a hot-film airflow sensor to better meter fuel mix for increased engine output at higher fuel efficiency. For model year 1996, Weissach introduced its Varioram induction system. This utilized variable length intake pipes and separate, differently tuned air intake systems, depending on medium or high engine speed. This system, capitalizing on the difference in air flow at various engine loads, eked another 12 horsepower out of the 3.6-liter engine, raising output from 270 horsepower SAE net to 282 SAE net at 6,300, and increasing torque from 243 to 250 ft-lb SAE net at 5,250 rpm.

To protect owners' investments starting in model year 1994, Porsche also introduced the "Electronic Immobilizer," called "Drive-Block," that acts on the Bosch engine management DME system.

To protect owners' exuberance beginning in model year 1995, Porsche had introduced a new Automatic Brake Differential (ABD) traction control system as an option. This automatically applies braking to the one rear wheel that is spinning during acceleration up to 44 miles per hour. This allows the driver to apply maximum engine torque to marginal road conditions. On six-speed manual transmission cars, the ABD is included with the optional limited-slip transmission, incorporating a 25 percent differential lockup on accelerating and 40 percent on deceleration. It was standard on the Carrera 4 models.

Where Typ 964 Carrera 4 models had relied on constant velocity joints and a complicated computer system to transmit power to the front tires, Porsche simplified the mechanism greatly by using a viscous center clutch, eliminating the sluggish front drive sensations. Porsche made good use of this improvement with its model-year 1996 turbo-bodied C4S coupe. (Porsche added a wide-bodied C2S coupe for mid–model year 1997, introduced in March.)

The Weissach rear suspension developed for the 928 models and adapted to the 964 Carrera 2 and 4 automobiles led to improvements in the rear suspension on this new Typ 933. A subframe carries both lower and upper A-arms with additional transverse links. This rear subframe and the front subframe are so strong that if these plates are bent in a crash, they cannot be straightened. They must be replaced.

Porsche gave enthusiast drivers a decidedly Formula One racing-inspired modification with its new Tiptronic S transmission, introduced mid-1995 model year. In order for drivers never to lift their hands from the steering wheel, Porsche was the first auto-maker to place the shifter mechanisms on the steering wheel hubs, connected electronically to the four-speed gearbox. Porsche retained the center tunnel-mounted shifter as well. The Tiptronic and Tiptronic S were available on two-wheel-drive cars only, but the S steering wheel controls could be refitted to earlier 1995 Tiptronic-equipped cars.

The Targa had returned to Porsche's line-up in 1996, but it was not without problems. However, virtually all the problems were addressed and remedied with the 1998 model year. The problems with the 1996 and 1997 model year cars are serious enough that it would be wise as a buyer to avoid these two models. One last note. Ferdinand (Ferry) Porsche Jr. who, with his father, had created a legendary sports car company, died on March 27, 1998. He was 88.

1998 Specifications "W" Series

Body Designation:		Typ 993 Carrera S coupe, Targa, Cabriolet, Carrera 4S coupe, Carrera 4 Cabriolet
Price:		911 Carrera S coupe: $63,750
		Cabriolet: $73,000
		Targa: $70,750
		911 Carrera 4S coupe: $73,000
		Carrera 4 Cabriolet: $78,350, add for Tiptronic: $ 3,265
Engine Displacement and Type:		911 Carrera S, 4S coupe: Typ M64/23; 3,608 cc, 219.9 cid, SOHC, Bosch Digital Motor Electronic (DME) management system, dual-spark, twin-plugs. (M64/24, as above, for Tiptronic models)
Maximum Horsepower @ rpm:		Carrera S, 4S: 282 SAE @ 6,300 rpm
Maximum Torque @ rpm:		Carrera S, 4S: 250 ft-lb @ 5,000 rpm
Weight:		Carrera S: 3,064 pounds
		Carrera 4S: 3,197 pounds
		Targa: 3,130 pounds, add 64 pounds for Tiptronic
0–60 mph:		Carrera S: 5.3 seconds (factory)
		Carrera 4S: 5.2 seconds (factory)
		Carrera Tiptronic: 6.3 seconds (factory)
Maximum Speed:		Carrera S: 171 mph (factory)
		Carrera 4S: 168 mph (factory)
		Carrera Tip: 168 mph (factory)
Brakes:		Carrera: power vented, cross drilled, dual-circuit disc brakes w/ABS 5 anti-locking brake system, four-piston aluminum alloy fixed caliper, cross-drilled
Steering:		Hydraulically assisted, force-sensitive rack-and-pinion
Suspension:	Front:	MacPherson struts with telescoping shock absorbers, lower control arms
	Rear:	Multilinks, self-stabilizing toe characteristics, antisway bar
Tires:	Front:	Carrera S, 4S: 205/50ZR17 optional: 225/40ZR18
	Rear:	Carrera S, 4S: 255/40ZR17 optional: 285/30ZR18
Tire air pressure:		All models, front and rear: 36 psi
Transmission(s):		Carrera S, 4S: G50/20 6-speed Tiptronic: A50/05
Wheels:	Front:	Carrera: 7.0Jx17 optional: 8.0Jx18
	Rear:	Carrera: 9.0Jx17 optional: 10.0Jx18

What they said at the time–Porsche for 1998

1998 Porsche 911, *Road & Track*, February 1998

By Kim Reynolds

"If you've ever spent a long night along California's twisty Highway 1, far up the coast, downshifting into fog-laden curves between pine trees with that air-cooled flat-six whooping behind you, feeling the steering tremble its Braille description of the pavement's surface, watching the close, no-nonsense instruments paint the backs of your fingers with their glow, well, you'd be passionate too. There are few cars left on this planet that can mesmerize."

Parts List for 1998 911s

These are items most commonly replaced during regular maintenance and routine daily operation. Prices quoted are for new factory parts at list price, not including installation labor. NLA means factory parts no longer available, so prices quoted are from aftermarket suppliers.

Engine:

1. Oil filter set.............. $54.30
2. Alternator belt set..... $57.14
3. Starter...................... $1,287.90
4. Alternator $1,035.80
5. Muffler .... $725.90 each
6. Clutch disc $284.56

Body:

7. Front bumper............ $1,666.11
8. Left front fender........ $2,641.01
9. Right rear quarter panel $3,309.82
10. Front deck lid $1,797.10
11. Front deck lid struts.. $37.23 each
12. Rear deck lid struts... $32.88 each
13. Porsche badge, front deck lid............ $142.29
14. Taillight housing and lens $258.18
15. Windshield with antenna $1,418.04
16. Windshield weather stripping.................... $109.50 set

Interior:

17. Dashboard................ $2,888.65
18. Shift knob $186.40
19. Interior carpet, complete $900.00

Chassis:

20. Front rotor set $506.50
21. Brake pads, front set $264.48
22. Rear shock absorber . $397.67
23. Front wheel $707.37
24. Rear wheel $1,012.22

Ratings

1998 models, manual transmission

	Carrera coupe S	Carrera Targa	Carrera 4 Cabrio	Carrera 4S
Acceleration	4.5	4.5	4.5	4.5
Comfort	4.5	4.5	4.5	4.5
Handling	4.5	3.5	4	5
Parts	4	4	4	4
Reliability	4.5	4	4.5	4.5

1998 models, manual transmission, continued (2)

	911 Turbo	911 Turbo S
Acceleration	5	5
Comfort	5	5
Handling	5	5
Parts	4	4
Reliability	5	5

1998 models, Tiptronic S transmission

	Carrera S coupe	Carrera Targa
Acceleration	4	4
Comfort	4.5	4.5
Handling	4.5	3.5
Parts	4	4
Reliability	4.5	4

1998 1/2 model 996, manual transmission and Tiptronic S

	Carrera coupe	Carrera Cabrio
Acceleration	4.5	4.5
Comfort	5	5
Handling	4.5	4
Parts	4	4
Reliability	4	4

1998 Garage Watch
Problems with (and improvements to) Porsche 911 models.

New Targa engineering. First generation roof welded on outside of body to reinforced convertible. Now welding to inside of reinforced coupe body after fixed roof removed. Much stiffer, quieter. Any lifting tendencies simply strengthen bond to car body.

Few changes. Porsche introduced next generation 911 with water cooling and another new body, the Typ 996, in Europe. In this last year of the air-cooled 911 engine, a series of problems, outlined here, were brewing for arrival when cars hit 20,000, 30,000, and 40,000 mile.

Windshield and rear window outer seals creaked. Windshield delaminates at antenna entry, as from 1995.

Steering rack leaks, boots tear, as from 1995.

Monroe shocks too soft, leak, as from 1995.

Front control arm bushings crack. Creates steering shimmy, as from 1995.

On-Board Diagnostics Generation Two provides four oxygen sensors, two catalytic converters. Valve guide leaks can foul sensors, killing sensor, catalytic converters, as from 1995. By the time Check Engine light appears, may be very serious, costly problem to solve.

A/C condenser-oil cooler fan resistors, relays, fail. A/C evaporator core leaks. A/C-heater computer/control unit fails, as from 1995.

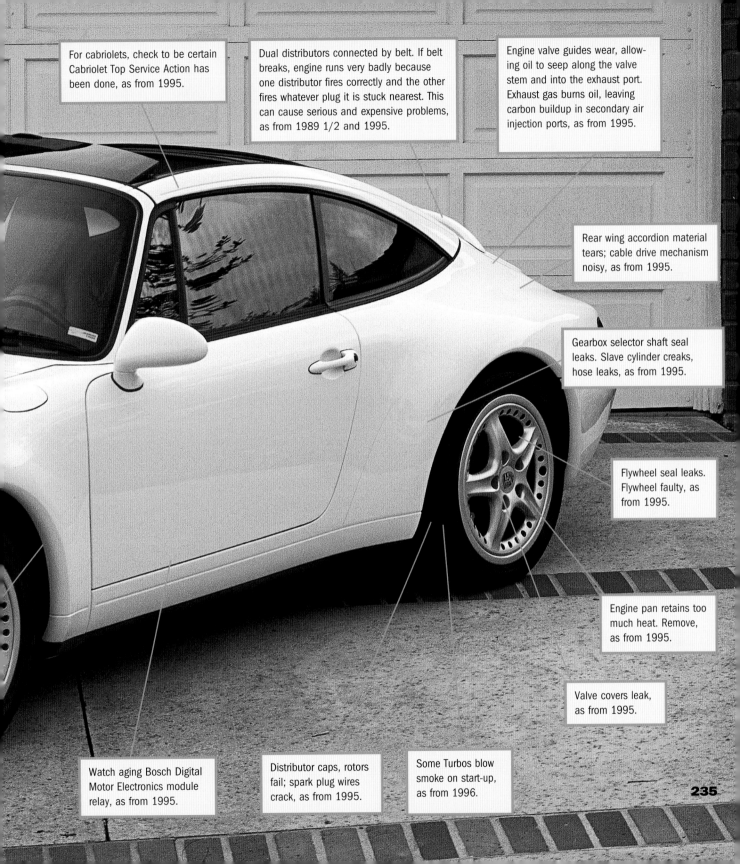

For cabriolets, check to be certain Cabriolet Top Service Action has been done, as from 1995.

Dual distributors connected by belt. If belt breaks, engine runs very badly because one distributor fires correctly and the other fires whatever plug it is stuck nearest. This can cause serious and expensive problems, as from 1989 1/2 and 1995.

Engine valve guides wear, allowing oil to seep along the valve stem and into the exhaust port. Exhaust gas burns oil, leaving carbon buildup in secondary air injection ports, as from 1995.

Rear wing accordion material tears; cable drive mechanism noisy, as from 1995.

Gearbox selector shaft seal leaks. Slave cylinder creaks, hose leaks, as from 1995.

Flywheel seal leaks. Flywheel faulty, as from 1995.

Engine pan retains too much heat. Remove, as from 1995.

Valve covers leak, as from 1995.

Watch aging Bosch Digital Motor Electronics module relay, as from 1995.

Distributor caps, rotors fail; spark plug wires crack, as from 1995.

Some Turbos blow smoke on start-up, as from 1996.

235

Index